T0219898

GEOGRAPHIES OF AUSTRALIAN HERITAGES

Heritage, Culture and Identity

Series Editor: Brian Graham,
School of Environmental Sciences, University of Ulster, UK

Geographies of
Australian Heritages
Loving a Sunburnt Country?

Edited by

ROY JONES
Curtin University of Technology, Perth, Western Australia

BRIAN J. SHAW
The University of Western Australia

Routledge
Taylor & Francis Group

LONDON AND NEW YORK

First published 2007 by Ashgate Publishing

Reissued 2018 by Routledge
2 Park Square, Milton Park, Abingdon, Oxon OX14 4RN
605 Third Avenue, New York, NY 10017

First issued in paperback 2021

Routledge is an imprint of the Taylor & Francis Group, an informa business

A Library of Congress record exists under LC control number: 2007005506

Notice:
Product or corporate names may be trademarks or registered trademarks, and are used only for identification and explanation without intent to infringe.

Publisher's Note
The publisher has gone to great lengths to ensure the quality of this reprint but points out that some imperfections in the original copies may be apparent.

Disclaimer
The publisher has made every effort to trace copyright holders and welcomes correspondence from those they have been unable to contact.

ISBN 13: 978-0-8153-8920-0 (hbk)
ISBN 13: 978-1-3511-5752-0 (ebk)
ISBN 13: 978-1-138-35698-6 (pbk)

DOI: 10.4324/9781351157520

Contents

List of Figures

List of Tables

Notes on Contributors

Graeme Aplin is an honorary Senior Research Fellow in the Department of Human Geography at Macquarie University, Sydney, having retired at the end of 2004. His research and writing have concentrated on the historical geography of Sydney, environmental studies with an Australian focus, and heritage issues. His publications include *Australians and Their Environment: An Introduction to Environmental Studies* (Melbourne, Oxford University Press, 1998/2000), *Heritage: Identification, Conservation and Management* (Melbourne, Oxford University Press 2002) and co-authored works *Waterfront Sydney 1860–1920* (Sydney, Allen & Unwin, 1984/1991) and *Global Environmental Crises: An Australian Perspective* (Melbourne, Oxford University Press, 1994/1999). He served as Editor of *Australian Geographer* from 1993 to 2000 inclusive, and is a member of Australia ICOMOS.

Nicholas Gill is a Lecturer in Geography at the University of Wollongong, NSW. His research into social and cultural aspects of pastoral land use in Central Australia has resulted in numerous publications on settler and indigenous pastoralism in Australian within international journals such as the *Journal of Rural Studies*, *Society and Natural Resources*, *Australian Humanities Review*, *Australian Archaeology*, *The Rangeland Journal* and *Australian Geographical Studies*, as well as in several edited collections. He has worked as an assistant curator at the National Museum of Australia where he collected and documented material culture from the inland pastoral industry.

Colin Michael Hall is a Professor in the Department of Management at the University of Canterbury, New Zealand. He also holds visiting positions at the University of Oulu in Finland and Lund University in Sweden. He is co-editor of *Current Issues in Tourism* and former chair of the IGU Commission on the Geography of Tourism, Leisure and Global Change. He is an author or editor of over 40 books and 200 other publications and his PhD on Australian wilderness heritage was subsequently published by the Melbourne University Press.

Marion Hercock has been a Company Director of Explorer Tours since 2001. She graduated with a BA (First Class Honours) and PhD from The University of Western Australia and was a Lecturer in Geography in the Faculty of Natural and Agricultural Sciences at UWA. Her research into environmental policy and management, island biogeography, environmental history, geographical thought, tourism and recreation has been published in a wide range of journals including *Australian Parks and Leisure*, *The Environmentalist*, *Geography* and *Geographers: Biobibliographical*

Studies. A member of the Royal Western Australian Historical Society, she is a reader and member of the *Expeditions of Exploration in Western Australia* project.

Colin Ingram is a manager in the Western Australian Department of Environment and Conservation, where he has responsibility for recreation and tourism policy and social research coordination. He has over 20 years of experience in park management in two Australian States. He has qualifications in environmental management from the University of Canberra and is currently a postgraduate research student at Curtin University of Technology.

Roy Jones is Professor of Geography, Dean of the Centre for Research and Graduate Studies in the Division of Humanities and Co-Director of the Sustainable Tourism Centre at Curtin University of Technology, where he has worked since 1970. He is an historical geographer with research interests in sustainability and heritage issues. He has been Human Geography Editor of 'Geographical Research: Journal of the Institute of Australian Geographers' since 2000.

Catherine Kennewell (nee McDonald) recently graduated from The University of Western Australia with Honours in Geography, and is currently working as a Planner with the WA Department for Planning and Infrastructure.

Andrew Kingham is a PhD student at Curtin University of Technology and author of 'Summary Report on Findings of Surveys of Unmanaged Camping in the North-West Cape Region of Western Australia' (2002). His background includes international tourism project management, research, teaching and management consultancy to government and the private sector.

William (Bill) Logan holds the UNESCO Chair in Heritage and Urbanism at Deakin University in Melbourne, Australia, and is director of Deakin's Cultural Heritage Centre for Asia and the Pacific. As well as a member of ICOM, he is a member of the Australia ICOMOS Executive Committee and was its national president from 1999 to 2002. He chaired the Australian National Cultural Heritage Forum, the federal minister's advisory committee, in 2002. His teaching, research and consultancy work focuses on cultural heritage conservation, especially relating to heritage theory, World Heritage, urban heritage, and heritage education and training. His recent publications include *The Disappearing Asian City: Protecting Asia's Urban Heritage in a Globalizing World* (Oxford University Press, 2002) and *Hanoi: Biography of a City* (UNSW Press, U. Washington Press & Select Publishing, 2000). He was a Visiting Fellow at ICCROM, Rome, in April 2003, and is a member of the International Board of the Academy of Irish Cultural Heritages at the University of Ulster.

Alistair Paterson is a Lecturer in Archaeology at The University of Western Australia. His research into cultural interaction between European settlers and Australian Aborigines in Central, Western and Northern Australia has resulted in regional historical archaeological studies of 19th and 20th century pastoral, mining and maritime industries as well as of transformations in indigenous and settler worlds. His research has been published in international journals such as *Australian Archaeology, Archaeology in Oceania, Historical Archaeology,* and *Australasian Historical Archaeology*, and in several edited collections. He is co-editor, with Jane Balme, of *Archaeology in Practice: A Student Guide to Archaeological Analyses* (Blackwell Publishing, 2005).

Matthew Rofe is Senior Lecturer in Planning at the University of South Australia. His research interests are in urban and cultural geography and his doctoral thesis was entitled *'I want to be global': theorising the gentrifying class as an emergent elite global community* which examined the restructuring of social structures and spatial networks under the contemporary influences of globalisation. He has also conducted research into residential segregation, gated communities, sexuality, masculinity and identity performance amongst sub-cultural groups.

Rosemary Rosario is a principal of the heritage architectural consultancy *Heritage and Conservation Professionals*, established in Perth, Western Australia in 1992. She graduated with a BArch (Hons) and MPhil (Urban Studies) from The University of Western Australia. During the 1970s she worked as a practicing architect in the UK. In her professional life she specialises in heritage assessment, conservation policy and general architectural and planning advice on heritage related issues. She has carried out numerous projects for state, local government and private clients. She was a member of the Heritage Council of Western Australia from 1998 to 2002.

Brian J. Shaw is Senior Lecturer in the School of Earth and Geographical Sciences at The University of Western Australia, Perth. His research into urban development, heritage and tourism issues has been widely published in journals such as *Australian Geographer, Current Issues in Tourism, GeoJournal, International Journal of Heritage Studies, Malaysian Journal of Tropical Geography* and *Urban Policy and Research*. His recent books include joint authorship of *Beyond the Port City: Development and Identity in C21st Singapore* (Pearson 2004) and co-editorship of *Challenging Sustainability: Urban Development and Change in Southeast Asia* (Marshall Cavendish 2005).

Wendy Shaw is a Lecturer in Geography at the University of New South Wales, Sydney. Her research within the field of Indigenous Geographies has been published in a number of book chapters and within international journals such as *Antipode, Australian Geographical Studies, Geografiska Annaler,* and *Gender, Place and Culture*. She is author of *Cities of Whiteness* (Blackwell, Oxford) and co-editor of a forthcoming volume *Indigenous Geographies: Critical Approaches*.

Hilary Winchester is the Pro Vice Chancellor and Vice President: Organisational Strategy and Change at the University of South Australia. She has taught a full range of undergraduate courses in human geography and supervised postgraduates at masters and doctoral level and her research interests are focused on key social issues such as urban poverty, population change, the geography of families and the impact of development.

Acknowledgements

The research work described in Chapter 8 was largely supported by the Australian Institute of Aboriginal and Torres Strait Islander Studies and Land and Water Australia. The authors also acknowledge the support of staff at the Central Land Council in Tennant Creek and the Arabana people for the conduct of the case studies.

The J.S. Battye Library of Western Australian History granted permission to reproduce the images of houses in City Beach in Chapter 10.

For First Generation Australians
Josephine and Bernard Shaw and Lee,
Ceridwen and Lewis Jones

Chapter 1

Introduction: Geographies of Australian Heritages

Roy Jones and Brian J. Shaw

Heritages: things worth saving?

As other geographers before us have acknowledged, heritage 'is an idea that is being increasingly loaded with so many different connotations as to be in danger of losing all meaning' (Graham et al. 2000, 1). The many ways in which heritage can be divided include: topically (natural and cultural – and a whole range of subsets, such as botanical and architectural, under these broad headings); ethnically (Indigenous, settler, migrant – and into numerous subgroups within these classifications); perceptually (tangible and intangible or, more broadly, experienced by the senses or the intellect); and by scale (local, national, global and various levels in between these three). Yet all these sub-categories of heritage will contain an array of items, which are considered to be, in Lowenthal's (1979, 555) words, 'things worth saving'. However, precisely what is considered to be worth saving will change over time. Graham et al's (2000, 2) claim that the 'key defining element' of heritage is 'the *present* (our emphasis) needs of the people' is significant. For example, in late nineteenth century Australia the 'convict stain' was something to be eradicated. By the late twentieth century, convict-built – and convict built – heritage was widely valorised and many Australians now search increasingly sophisticated databases in search of convict ancestors.

Thus Graham et al. remind us that heritage is fundamentally a contemporary phenomenon. While we may wish to save things from the past for the future, it is the opinions, decisions and actions of people in the present that bring about their salvation and preservation or, indeed, their obliteration. Furthermore, both Graham et al (2000, 3) and Lowenthal, through his use of the term 'worth', remind us of the significant and growing economic dimension of heritage. Whatever their scientific, aesthetic or sociocultural merit, it is their financial 'worth' that is an increasingly important factor in decisions on whether particular heritage items are presently 'needed'.

But, while the commercialisation and, thus, the commodification of heritage is one 'connotation' by which the term can lose elements of its meaning, a further problem arises when the question 'Who are the "people" whose present needs are to be met?' is put. In settler and multicultural societies, such as Australia, the views of different groups on which 'things' are and are not 'worth saving' in this context are likely to diverge considerably.

Australian heritages

For most of Australia's colonial and postcolonial history, its various governments strove to impose a British heritage on the continent and its inhabitants. In many ways, their enduring success in this regard has been impressive. The English language is the undisputed means of both official and general communication for the vast majority of the population. As late as 1952, an article in the *Adelaide Advertiser* referred to 'English' and 'foreign' migrants as separate categories (Jones 1995, 251) and strong discrimination in favour of, if not English, then certainly British and, to a lesser extent, European, migrants continued until the 1970s. And, in 1999, Australians voted to retain a British head of state.

But other heritages either remain or emerge. Given both its size and its location, the natural environments of Australia are diverse, globally idiosyncratic and radically different from those of the British Isles. For the settler groups, as for the Indigenous population that preceded them, the colonisation of Australia involved coming to terms with a challenging and initially unfamiliar physical environment (Bolton 1981) and, for at least some of these settlers, this has also entailed an identification with the continent's natural landscapes and an acknowledgment that elements of these landscapes are 'worth saving' (Powell 2005). Indeed Dorothea McKellar made this identity shift as early as 1906

> The love of field and coppice,
> Of green and shaded Lanes,
> Of ordered woods and gardens,
> Is running in your veins;
> Strong love of grey-blue distance,
> Brown streams and soft, dim skies –
> I know but cannot share it,
> My love is otherwise.
>
> I love a sunburnt country,
> A land of sweeping plains,
> Of ragged mountain ranges,
> Of drought and flooding rains,
> I love her far horizons,
> I love her jewel sea,
> Her beauty and her terror –
> The wide brown land for me.

Just as these landscapes could not be 'embellished' (Stannage 1990) until they became something that was merely British, so the Indigenous population could not be simply obliterated. While the debate over the attempt to wipe out Tasmania's Aboriginal population continues (Macintyre 2004; Windschuttle 2002) the early and even mid twentieth century belief that the role of the colonisers was merely to 'smooth the dying pillow' (Bates 1938) of the Indigenous population was widespread. But, in the early twenty first century, the Aboriginal population has once more reached pre-conquest levels, and Indigenous birth rates are significantly higher than those of virtually all the settler populations. In practical terms, serious levels of Indigenous disadvantage remain, notably, but certainly not only, in health standards and life expectancy rates. But, in comparison with the assumption of 'terra nullius' by the first European colonisers, the upholding of the existence of a form of Native Title in the Mabo and Wik High Court decisions of the 1990s is at least a step forward in the acknowledgment of Australia's Indigenous heritages.

From its (British) birth in 1788, there has also, and always, been cultural and, thus, heritage diversity within Australia's settler population. Initially, the Irish were the most culturally significant non-British group (Proudfoot 2003). Although Ireland was part of the United Kingdom from 1800 to 1922, this group always maintained a (largely Catholic) identity separate from that of the predominantly Protestant British Australians. Both Ned Kelly, the bushranger, in his Jerilderie letter of 1879, and Peter Lalor, the leader of Eureka Stockade rebellion in 1854, invoked their Irish heritage in their opposition to British control of Australia. But Lalor's lieutenants included a Prussian and an Italian and the gold rush which led to the Eureka rebellion also precipitated the arrival in Australia of large numbers of 'diggers' from all over the world, though it was the influx of the Chinese that gave rise to most contemporary concern.

Over the course of the second half of the twentieth century successive waves of migrants added to the country's cultural diversity. These settlers came from Southern and Central Europe in the immediate post-war period and subsequently from the Eastern Mediterranean, East and South East Asia and the Pacific. The proportion of overseas-born in the population fell from 23 per cent in 1901 to 10 per cent in 1947, but then rose to 24 per cent by 2004. Significantly perhaps, this early twentieth century fall in the proportion of overseas born coincided with Australia's first decades as a unified (albeit federal) and (relatively) independent nation. Patriotism and nationalism were fostered both by the progress towards federation at the end of the nineteenth century and by the country's participation in two World Wars. By mid-century, therefore, a distinctive Australian identity and heritage were developing in a way that made the country 'very familiar and awfully strange' (Hammerton and Thomson 2005, 124) to the Ten Pound Poms arriving there from Britain. In subsequent decades an increasingly diverse nation has sought to define and redefine a distinctively Australian heritage and identity in which this diversity has been variously celebrated, as in the endorsement of multiculturalism in the 1970s, and decried, as in the contemporary search for a single set of 'Australian values'.

Geographies of Australian heritages

To paraphrase Graham et al. (2000, 4–5), heritage occurs in place(s), it is important to people (s), it is inevitably 'context bound and power laden' and it is 'an economic good and is commodified as such'. It therefore mirrors geography in its concerns for places/environments, for peoples and their identities, for power and conflict (particularly over space) and for local and regional development. Given these multiple similarities, it is not surprising that heritage issues arise in a wide range of geographical sub fields from environmental management through cultural, historical, political and economic geography to urban and regional development and tourism.

With this in mind, the remaining chapters of this volume provide a series of geographical perspectives on a selection of Australian heritages. In compiling this collection we have sought to mix the environmental and the cultural, the metropolitan and the remote, the economic and the political and the academic and the applied by bringing together a group of authors who are predominantly geographers by training, though their current academic positions also encompass Planning, Tourism, Archaeology, Heritage and Urbanism and University Management. We have also moved beyond academia to include contributors working in Park Management and Heritage Tourism as well as a private Heritage Consultant. In doing so, we have sought to illustrate some of the varied ways in which inhabitants of this sunburnt country have chosen – or have not chosen – to love it.

Loving a sunburnt country?

In Chapter 2, Graeme Aplin carefully outlines the legislative and bureaucratic framework for heritage protection in Australia. He details the responsibilities of the federal government and those of the respective states, territories and local authorities, while also considering the roles of voluntary and community organisations. While making the point that the bulk of practical, day-to-day heritage responsibility rests with the states and territories, or with local government, he devotes particular attention to the recent overhaul of federal environment and heritage legislation by the Howard Coalition Government, an issue that is later revisited by William Logan in the concluding chapter.

In Chapter 3, Aplin extends this introductory analysis in the global context, specifying the part played by Australia under the terms of the World Heritage Convention. He provides a detailed analysis of Australia's sixteen World Heritage properties and the contenders for possible future nomination, detailing the sometimes-thorny relationships that exist between the Australian Government and World Heritage bodies in relation to controversial development issues in places such as Kakadu National Park. The focus on National Parks is continued in Chapter 4 where C. Michael Hall documents the changing geographies of Australia's wilderness heritage. This chapter discusses the wilderness concept and provides an insightful historical account of the Australian wilderness conservation ethic and movement. Hall also details the growth in Commonwealth regulatory capacity and state responsibilities, together with the ongoing importance of tourism as an economic rationale for wilderness conservation.

In Chapter 5, Marion Hercock reveals a different perspective on wilderness heritage, writing on the ABC of running an innovative heritage-based tourism operation. Based on her own personal experiences, she details the wider economic and social context of global finance and markets that impinge upon her operation; the 'paradox of place and places' which adds to the problem of marketing little-known sites; and some of the 'on the ground' complexities created by the local social setting, which includes government administration of conservation reserves, the private management of mining and pastoral leases and the administration of Aboriginal lands, as well as the physical environment and the unpredictability created by rain or storm events. This essay raises some interesting questions regarding the sustainability of small heritage tourism operations.

Chapter 6, written by Roy Jones, Colin Ingram and Andrew Kingham, demonstrates how the three central characters of Australia's iconic Waltzing Matilda narrative, namely 'swagmen', 'squatters' and 'troopers', intersect in today's 'outback' area of the North West Cape – Ningaloo Reef region of Western Australia. Their present-day equivalents, being unauthorised wilderness campers, production-oriented pastoralists and local regulatory authorities, are now increasingly engaged in a number of contestations regarding access to key resources, the preservation of natural heritage and the management of the rapidly growing tourism industry.

In Chapter 7, Wendy Shaw draws our attention to notions of tradition and heritage as they are applied to Indigenous peoples in Australia. She argues that because Aboriginal people and their associated places have been disengaged from mainstream experiences, they have become museum-like objects. Consequently Indigenous heritage remains discursively locked in archaeological pasts, and urbanised Indigeneity in places such as *The Block*, in the inner Sydney suburb of Redfern, is constructed as 'out of place'. She compares this contemporary spatial reality to Sydney's Rocks area that has now been (re)fabricated and cleansed to represent an idealised and sanitised history of colonisation. This theme of deliberate demarcation between indigenous and settler heritage is continued in Chapter 8 where Nicholas Gill and Alistair Paterson write about Aboriginal people and Australian pastoral cultural heritage, whereby the myriad Aboriginal involvements in pastoralism have been largely forgotten and have gone unmarked. The authors emphasise the diversity of Aboriginal associations with pastoralism and pastoralists, the variety of Aboriginal pastoral landscapes, and the continued relevance and dynamism of Aboriginal associations with pastoralism, through a number of case studies utilising archaeological and geographical perspectives.

Chapter 9 provides us with an invaluable insight into South Australia's long and distinctive Germanic heritage, set in the context of rural idyll place making. Matthew W. Rofe and Hilary P.M. Winchester trace the changing nature and representations of Germanic heritage and reveal the contested nature of local place identity in the Adelaide Hills village of Lobethal, established by German Lutherans in 1842. The village's contemporary construction as a Christmas wonderland has reinvented and reinvigorated a declining rural community but the authors look beyond the obvious boosterism to unravel the complexity inherent in the landscape and the highly subjective nature of heritage place making.

The remaining case studies deal with urban-based heritage in Perth, Western Australia. In Chapter 10, Catherine Kennewell and Brian J. Shaw examine Perth's 1962 Commonwealth Games legacy. More than 'just' sporting heritage, the Games were instrumental in exposing the rather parochial capital of Australia's hitherto 'Cinderella State' to the much wider world, yet the future prospects for Perry Lakes stadium, the main venue for track and field events, and the award-winning Games Village in City Beach, are now decidedly uncertain. The authors trace the demise of these structures in the context of vested residential interests, urban planning initiatives and the broader issues that relate to the upgrading and relocation of sporting stadia. The realisation that any objective assessment of heritage value quickly disappears in the face of potentially reduced resale values, or of restrictive development guidelines on private properties, is a recurring theme that resurfaces in Rosario's chapter.

Roy Jones, who writes on Port, Sport and Heritage in Western Australia's historic port city of Fremantle, continues the sporting theme in Chapter 11. While the value of Fremantle's built heritage is unquestioned, the lack of National Heritage Listing notwithstanding, the city has other significant roles, notably as a major port and as a significant service, entertainment and tourist centre. Jones traces the history of Fremantle's development and tells the story of how these roles have become increasingly intertwined and, on occasions, contested, in a climate of economic, demographic, social and cultural changes. Contestation is also the dominant theme underpinning Rosemary Rosario's Chapter 12, entitled 'Places Worth Keeping', which considers the case for the protection of heritage at the local level. Written from the perspective of a heritage professional, this chapter reviews some the issues relating to the City of Subiaco's release of the review of its municipal inventory, looks at their immediate aftermath, and some longer-term consequences. It concludes by addressing the ways in which local heritage can be managed successfully, calling for the heritage message to be clearer, easier to understand, more consistent and, above all, balanced.

Finally, Chapter 13, written by William S. Logan, provides a reflective summary of the issues encapsulated in the wider volume. This concluding chapter re-casts heritage as an element of Australian cultural politics, exploring the linkages between ideology and conservation practice. It considers some of the current difficulties being experienced by the Australian heritage system that render it vulnerable to political exploitation. Moreover, Logan echoes some of the concerns expressed by the heritage professionals writing in this volume, that the system is over-extended in respect to its planning control functions, while still narrow in heritage content and fragmented in its efforts to provide appropriate legislation and administration. Most appropriately, in a volume such as this, he concludes by recommending a special role for geographers, whereby their unique set of synthesising skills and interests can be marshalled to challenge the neo-liberal social and economic development approach. The geography agenda insofar as it relates to heritage issues can embrace the intangible values of places and help to achieve more holistic and culturally sensitive approaches to environmental understanding and protection.

References

Bates, D. (1938) *The Passing of the Aborigines: A Lifetime Spent among the Aborigines of Australia* (London: Murray).

Bolton, G. (1981) *Spoils and Spoilers: Australians Make their Environment 1788–1980* (Sydney: George Allen and Unwin).

Graham, B., Ashworth, G. and Tunbridge, J. (2000) *A Geography of Heritage: Power, Culture and Economy* (London: Arnold).

Hammerton, J. and Thomson, A. (2005) *Ten Pound Poms: Australia's Invisible Migrants* (Manchester: Manchester University Press).

Jones, R. (1995) 'Far Cities and Silver Countries: Migration to Australia in Fiction and Film', in King et al. (eds).

King, R., Connell, J. and White, P. (eds) (1995) *Writing across Worlds: Literature and Migration* (London and New York: Routledge).

Lowenthal, D. (1979) 'Environmental Perception: Preserving the Past', *Progress in Human Geography* 3:4, 550–559.

MacIntyre, S. and Clark, A. (2004) *The History Wars* new ed. (Carlton: Melbourne University Press).

Powell, J. (2005) 'Environment-Identity Convergences in Australia, 1880–1950' in Proudfoot and Roche (eds).

Proudfoot, L. (2003) 'Landscape, Place and Memory: Towards a Geography of Irish Identities in Colonial Australia', in Walsh (ed.).

Proudfoot, L. and Roche, M. (eds) (2005) *(Dis)Placing Empire: Renegotiating British Colonial Geographies* (Aldershot: Ashgate).

Stannage, T. (1990) *Embellishing the Landscape: The Images of Amy Heap and Fred Flood* (Fremantle: Fremantle Arts Centre Press).

Walsh, O. (ed.) (2003) *Ireland Abroad. Politics and Professions in the Nineteenth Century* (Dublin: Four Courts Press).

Windschuttle, K. (2002) *The Fabrication of Aboriginal History: Vol.1, van Diemen's Land 1803–1847* (Paddington, N.S.W.: Macleay Press).

Chapter 2

Heritage Protection in Australia: The Legislative and Bureaucratic Framework

Graeme Aplin

A Federation

Australia is a federation with three levels of government – federal, state and local.[1] The responsibilities of the Federal Government, on the one hand, and of the eight states and territories, on the other, are defined by the Constitution. Consequently, the role of the Federal Government in heritage matters is limited, and that of the states relatively extensive. Federal responsibilities are limited to those either specified in the Constitution or agreed to by the states and, in practice, centre on matters of national importance or related to international agreements, especially the World Heritage Convention (see Chapter 3). Although details vary from state to state, local or municipal government also has an extremely important part to play because of the planning powers it holds. Hence, the bulk of practical, day-to-day heritage responsibility rests with the states, or with local government within the states.

Further background

Australia has a long history of human occupation, the Aborigines having arrived at least 60,000 years ago. Non-indigenous Australians long thought the Indigenous peoples had left few permanent signs of their occupation, and hence had little cultural heritage of importance, but there has recently been an increased recognition of the richness of that heritage, stemming from the great variety that exists among Indigenous lifestyles and cultures, and from the intimate and spiritual relationship of the Indigenous peoples to their country.[2] On the other hand, permanent non-indigenous settlement only began in 1788. In fact, much of the continent was first settled by Europeans somewhat later, in some places not until the twentieth century. So for people with a strongly ingrained European view of 'cultural heritage', Australia has little or nothing pre-dating the start of the nineteenth century, and thus little cultural heritage to offer, especially if one believes that something has to be old to have real value.

---

1   Purely for convenience, the states and territories collectively will be referred to as 'the states'.

2   'Country' is the term used by the Indigenous peoples of Australia when referring to the environments and landscapes in which they live and to which they relate so strongly.

Fortunately, the general perception of heritage by Australians changed in the second half of the twentieth century (Aplin 2001, 183). Many people began to appreciate that even if European-Australian heritage is not very old, it is nevertheless very important. At the same time, many people recognised that there are important non-Anglo-Celtic and even non-European aspects to our non-indigenous heritage, aspects that have arisen from the increasingly multicultural and migrant-based nature of Australia's population. Furthermore, much greater recognition has been accorded Indigenous heritage, largely based on increased knowledge and appreciation, and a realisation that we have our own complex and fascinating Indigenous cultures in Australia, on a par with others anywhere. There also has been an increasing willingness to accept that inclusion of Indigenous heritage, as part of our national heritage, is crucial to reconciliation between Indigenous and non-indigenous Australians.

Australia, as perceived by its European settlers, is a young country that has developed very rapidly. The major focus has been on growth and change, as epitomised in the philosophies of *developmentalism* and *economic rationalism*, not on preservation of either natural landscapes or cultural heritage items, and gains in the heritage field have been hard won. In part, this is because of difficulties in giving economic value to intangibles such as heritage significance so that they can be incorporated on equal terms in benefit-cost analysis or environmental impact assessment.

The earliest activities in Australia that related to heritage, broadly defined, were concerned with conservation of natural areas, such as forests, and were more resource management than heritage management. Non-indigenous cultural heritage only became a political issue in the third quarter of the twentieth century, while a general community awareness of Indigenous heritage came even later. Although administrative structures vary between jurisdictions, as shown below, some general comments are relevant. First, natural heritage is commonly treated separately from cultural heritage, especially non-indigenous cultural heritage. When agencies were established to administer national parks, their concern was normally restricted to natural heritage, although they have, in fact, generally successfully managed non-indigenous and Indigenous cultural heritage items that happen to fall within park boundaries. However, the Australian definition of 'national park' is strongly based on natural heritage and biodiversity conservation, and park boundaries are usually delineated to exclude most 'modern' human occupancy. Australian national parks are thus generally quite unlike European ones.

When, more recently, non-indigenous cultural heritage was also seen as important, separate agencies were generally established, commonly in different departments and under different ministers. That was not true at federal level, however, and this was largely a reflection of the Federal Government becoming involved in heritage matters later than the states. At the state level, management of built heritage has most often been closely connected to planning, with much responsibility devolved to local government. Finally, for complex historical and philosophical reasons, the administrative 'home' for Indigenous heritage varies markedly between jurisdictions, and has done so within some jurisdictions over time.

The Federal Government

The Federal Government's heritage role is limited to areas involving its external or foreign affairs responsibilities, its own property and activities, and matters delegated to it by the states. In recent years it has also attempted to co-ordinate responses and promote uniform national standards and procedures, but this can only succeed with the co-operation and, ultimately, agreement of all states and territories. This degree of consensus has been difficult to achieve, as the nine governments have rarely all been of the same political persuasion, sharing the same philosophies. Furthermore, states have consistently guarded their own functions and responsibilities, and the concept of states' rights more generally.

The Federal Government, as the national government, has necessarily been the body negotiating, agreeing to, signing, and ratifying international agreements, including the World Heritage Convention (Chapter 3). Once any agreement is ratified, the Commonwealth must legislate to ensure that its conditions are met in Australia, but state legislation and other arrangements have also been needed on occasions, sometimes leading to major difficulties. Other foreign affairs or trade matters can also have heritage implications, and again the Commonwealth must become involved. The Federal Government is also responsible for its own heritage standards in relation to Commonwealth land and buildings, including those in the ownership or use of Commonwealth instrumentalities, and to actions of federal bodies that impinge on heritage values. In addition, following a 1967 referendum, the Federal Government has a relatively important role to play in Indigenous heritage issues.

Environment Australia

The Australian Government Department of the Environment and Heritage (DEH)) includes the Heritage Division (HD), which in turn supports the Australian Heritage Council (AHC). The Approvals and Wildlife Division and Parks Australia, both also within DEH, are important for natural heritage matters. The HD deals specifically with the management of Australian World Heritage properties (Chapter 3), as well as with other international treaty obligations, especially those concerning the Ramsar Convention, and the Convention on Biological Diversity (CBD), both of which have clear natural heritage relevance.[3] Despite having a relatively settled 'home' in recent years, heritage matters were shunted around from one department to another, and one minister to another, for decades, reflecting the relatively low priority given these matters by Australian voters, and hence politicians.

Australian Heritage Commission and Register of the National Estate

The Australian Heritage Commission was established in 1976, following the Hope Inquiry into the National Estate (which commenced in 1973). Constituted under

---

3   The Ramsar Convention – more correctly the Convention on Wetlands of International Importance Especially as Waterfowl Habitat – came into force in 1971, while the Convention on Biological Diversity was finalised in 1992.

the *Australian Heritage Commission Act (Cwlth) 1975*, the Commission's main aim was to 'identify, conserve and promote Australia's National Estate – those parts of the natural and cultural environment that have special value for current and future generations' (Australian Heritage Commission 1997, 4). It was required to prepare a Register of the National Estate (RNE), which was to be 'an inventory of all those parts of Australia's natural, historic, Aboriginal and Torres Strait Islander heritage which have special value for present and future generations ...' (Australian Heritage Commission 1997, 16–18). Among other things, listing places on the RNE provided information about their heritage value to assist in decision-making; obliged the Federal Government to consult with the Commission before taking any action that could harm or affect them; made their owners eligible for tax rebates for conservation works; and made them eligible for grants for identification, conservation or promotion under the National Estates Grants Program. On the other hand, listing places on the RNE did not give the Commission or the Commonwealth the right to acquire, manage or enter properties, or to restrict the activities of any non-Commonwealth person or entity. Listing was thus largely advisory, giving moral rather than legal protection, except in very limited circumstances stemming directly from the Commonwealth's powers. The number of listings was very large, as shown in the massive illustrated list of 6,600 places published after just five years of operation (Australian Heritage Commission 1981), and the more than 13,000 places listed by 2003.

New legislation, 1998–2003

Federal environment and heritage legislation was completely overhauled between 1998 and 2003, due to the Federal (Liberal Party and National Party Coalition) Government's desire to rationalise it and give as much responsibility as possible to the states. First, the *Aboriginal and Torres Strait Islander Heritage Protection Act (Cwlth) 1998* replaced an earlier Act in an attempt to reduce duplication between jurisdictions, largely through accreditation of state legislation. This Act allows for intervention where protection of objects and places is in the national interest. Secondly, the *Environment Protection and Biodiversity Conservation Act (Cwlth) 1999* (EPBCA) replaced a number of earlier acts in the general environment field, and included sections relating to heritage of national significance. These sections dealt with properties listed under international agreements, with listed threatened species and communities, with listed migratory species, and with the marine environment. Chapter 5, Part 15 of the Act is concerned with protected areas, and specifically with management of World Heritage properties, wetlands of international importance (Ramsar properties), and biosphere reserves (listed under the CBD). Various enforcement roles relate to such properties. Thirdly, bills specifically dealing with more general heritage issues were introduced into Federal Parliament in late 2000, and finally came into force in September 2003 as the *Environment and Heritage Legislation Amendment Act (No. 1) (Cwlth) 2003*, the *Australian Heritage Council Act (Cwlth) 2003*, and the *Australian Heritage Council (Consequential and Transitional Provisions) Act (Cwlth) 2003*.

The Government achieved its goal of reforming its heritage protection structures 'to achieve more effective protection of places of truly national importance', but the impact on other forms of heritage remains less certain. The legislation includes provision for a National Heritage List (NHL), a Commonwealth Heritage List (CHL), and an independent expert advisory body, the Australian Heritage Council (AHC), replacing the Commission. The NHL partially replaces the RNE, but is very much smaller, concentrating solely on places of 'outstanding national heritage significance'. The CHL is concerned specifically with places owned or managed by the Commonwealth and its agencies. One concern is that the NHL is very much leaner than the RNE (having just 34 places listed in late 2006),[4] with many places listed on the RNE now relying on protection under state jurisdictions. But then the RNE had little legal backing, as we have seen, and at least those places on the NHL do have the backing of the EPBCA for their protection, as discussed below. The RNE is retained, but solely as a database of heritage items, and for protective purposes many places on the RNE, but not the NHL, will be transferred to state and territory lists.

The AHC advises on the eligibility of places for listing on the NHL or CHL, and provides statements of significance, with the Minister deciding whether or not a place is finally listed. The Minister can also provisionally list a place if urgent protection is needed. On the other hand, the AHC itself can draw up management plans only for places owned by the Commonwealth. The AHC will also seek to influence state, territory and local governments to actively fulfil their heritage obligations, and will seek bilateral agreements where appropriate. In essence, places on the NHL are now included as an additional 'matter of environmental significance' under Chapter 5 of the EPBCA, meaning that the Minister must approve any activities that will significantly impact on their heritage values. The AHC can also provide technical and financial assistance for the preparation of management plans for places on the NHL, for their protection or conservation, and for their promotion, identification

---

4   As of late 2006, there were just 34 places listed on the NHL, although at least 60 had been nominated. Those listed were: Bud Bim National Heritage Landscape – Tyrendarra Area, SW Vic.; Budj Bim National Heritage Landscape – Mt Eccles Lake Condah Area, SW Vic.; Royal Exhibition Building National Historic Place, Melbourne, Vic.; Dinosaur Stampede National Monument, Central Qld; Kurnell Peninsula, Botany Bay, NSW; Eureka Stockade Gardens, Ballarat, Vic.; Castlemaine Diggings National Heritage Park, Vic.; Mawson's Huts and Mawson's Huts Historic Site, Antarctica; Brewarrina Aboriginal Fish Traps (Baiames Ngunnhu), NSW; Port Arthur Historic Site, Tas.; Glenrowan Heritage Precinct, Vic.; Sydney Opera House, NSW; Fremantle Prison, WA; First Government House Site, Sydney, NSW; Newman College, Melbourne, Vic.; Sidney Myer Music Bowl, Melbourne, Vic.; ICI Building (former)/Orica House, Melbourne, Vic.; Australian Academy of Science Building, Canberra, ACT; Recherche Bay (North East Peninsula) Area, Tas.; Richmond Bridge, Tas.; HMVS Cerberus, Port Philip Bay, Vic.; Melbourne Cricket Ground, Vic.; South Australia's Parliament House, Adelaide, SA; Tree of Knowledge, Barcaldine, Qld; Dirk Hartog Landing Site 1616 – Cape Inscription Area, WA; Batavia Shipwreck Site and Survivor Camps Area – Houtman Abrolhos, WA; Hermannsburg Historic Precinct, NT; The Australian War Memorial and Anzac Parade, Canberra, ACT; North Head, Sydney; Point Nepean Defence Sites and Quarantine Station, Victoria; Old Parliament House and Curtilage, ACT; Glass House Mountains Natural Landscape, Queensland; Rippon Lea House and gardens, Melbourne Vic.; Flemington Racecourse, Melbourne, Vic.

and presentation. Voluntary conservation agreements will enlist the help and co-operation of private owners. In addition, the Council will perform a monitoring role, with the Minister formally reporting to Parliament every two years. It appears that the Council will also co-ordinate a national database containing information on all heritage properties listed on the NHL, the CHL, and the various state, territory and local lists. In summary, there is potential for the new legislation to strengthen heritage protection in Australia, but only if the states and territories are prepared to play a genuine complementary role – and they appear to be prepared to do so, as discussed in the remainder of this chapter.

States and Territories

All state and territory governments have departments, divisions or agencies dealing with heritage, though their names and functions differ, as does their degree of independence from political interference. In all cases, governments and ministers retain the right to overrule departments and agencies if major projects are rejected, or major constraints are placed on development activities, when these projects, developments and activities are deemed to be in the state or public interest. In practice, this frequently means that economic considerations outweigh heritage ones, as is also the case at the federal level.

The following sections discuss the heritage framework in the six states and two territories. Although each has its own distinct heritage system, there are many common elements. All jurisdictions have agencies and legislation relating to both built (or European cultural) heritage and natural heritage, but it is probably in the treatment of Indigenous heritage that differences are greatest. Most of the heritage systems, especially as related to cultural or built heritage, have been developed only quite recently. However, this is rarely as recent as the dates of listed legislation indicate, as there has been on-going amendment and replacement of earlier Acts. Some common elements are discussed immediately below, and the sections that follow then concentrate on distinct aspects and recent and current developments in individual jurisdictions. Only the briefest of introductions can be provided, and other less directly relevant legislation, such as general planning legislation and specific wildlife and biodiversity conservation legislation, is not included. (For more detailed information on individual states and territories see websites listed in References).

General aspects

All states have a Heritage Council and maintain a Heritage Register, although exact titles differ, and a separate unit often provides administrative support. The Register usually includes heritage of all three major types. Procedures for listing vary, but generally allowance is made for public consultation and possible objection, and places may be entered on an interim basis prior to permanent listing. All states specifically involve local government in listing and protecting heritage sites of local, rather than state, significance, although, again, details vary. Most, if not all, states have provisions for government or ministerial intervention if an item is clearly threatened and urgent protection is needed, while all states provide advice and technical assistance to owners of listed sites and to other agencies, including local government, and sometimes financial assistance to owners (see Chapter 11). State agencies usually need to approve any works affecting listed properties, although this role is devolved to local government for properties of local significance. In many cases, however, the provisions mentioned above apply only, or most obviously, to non-indigenous cultural or built heritage.

Natural heritage is almost always administered separately, commonly by a National Parks Service (again, the name varies), while wilderness areas can be declared within, and sometimes outside, national parks and similar reserves. The treatment of Indigenous heritage is much more variable, but in most, if not all, cases, any Indigenous heritage finds anywhere in the state must be reported for recording and investigation, and Indigenous organisations in all cases have a vital role to play.

Australian Capital Territory (ACT)

The *Heritage Act (ACT) 2004* replaced earlier acts to 'provide for the recognition, registration and conservation of places and objects of natural and cultural significance'. Its provisions are administered by ACT Heritage, which also supports the ACT Heritage Council and administers the Heritage Grants Program. The new Heritage Register includes all types of heritage places, and is more comprehensive than its predecessors. Amendments to the *Land (Planning and Environment) Act (ACT) 1991* ensure integration of heritage matters into the development assessment process. Public land is reserved for a number of purposes, including national parks and nature reserves, and is administered by the Bush Parks and Reserves section of Environment ACT.

New South Wales (NSW)

Built heritage, the immoveable component of non-indigenous cultural heritage, is administered through the Heritage Council and its administrative arm, the Heritage Office (HO). The Council is constituted by, and administers, the *Heritage Act (NSW) 1977*. The HO administers the NSW Heritage Database, which includes all types of heritage significant at all scales, and the State Heritage Register (SHR), which is concerned only with heritage of state significance. While the Council has an Aboriginal Heritage Committee, it has no equivalent for natural heritage. Built heritage management, however, most frequently occurs through the protection of items by a local Council using its planning powers under a Local Environment Plan (LEP), or through Development Applications and Building Applications. The concerns of the Council and Office extend well beyond individual items, as typified by the initiative on cultural landscapes surrounding urban areas (NSW Heritage Office 2004).

Primary responsibility for both natural and Indigenous heritage lies with the National Parks and Wildlife Service (NPWS). The Service is responsible for national parks and other reserves, while its jurisdiction over Indigenous heritage extends well beyond such reserves to include all Indigenous heritage in the State. This means that any 'discoveries' of Indigenous sites, regardless of their degree of significance, must be reported to the NPWS, possibly for inclusion in its Aboriginal Heritage Inventory Management System.

Northern Territory (NT)

The *Heritage Conservation Act (NT) 1991* is managed by Heritage Conservation Services (HCS), which supports the Heritage Advisory Council, provides advice, and assists owners through Heritage Incentives. The scope of the Act encompasses natural and Aboriginal, as well as non-indigenous cultural, heritage. Under the Act, the Minister declares heritage places on the recommendation of the Council, while works affecting declared places must be in accordance with a conservation management plan prepared by the Council and approved by Parliament. The Act appeared to have been weakened in 1998, when amendments gave the responsible Minister greater powers to revoke a heritage order, or to authorise works, including alteration or demolition of declared places, regardless of management plans in force. This seemed to make heritage orders very exposed to the exercise of political whim, and to the developmentalist ethic that has dominated local politics. In early 2006, consultations were underway as part of a process to review the Act, partly because of what had seemingly become unduly and frequently cumbersome ministerial and parliamentary involvement.

National Parks (other than Kakadu and Uluṟu–Kata Tjuṯa, which are administered by Parks Australia) and other reserves are administered and managed by the Parks and Wildlife Commission according to the *Territory Parks and Wildlife Conservation Act (NT) 1976*. Management normally involves Traditional (Indigenous) Owners in major ways. Although HCS is concerned with Indigenous heritage and maintains the NT Archaeological Resources Database, other acts are also crucially important. The *Aboriginal Land Rights (Northern Territory) Act (Cwlth) 1976* protects sites that are 'sacred to Aboriginals or otherwise of significance according to Aboriginal tradition'. It is an offence for people other than Traditional Owners to enter, remain on, or carry out work on such a site without approval. The *Aboriginal Sacred Sites Act (NT) 1989* charges the Aboriginal Areas Protection Authority with maintaining a Register of Sacred Sites, which have similar restrictions attached to them.

Queensland (Qld)

The scope of the *Heritage Act (Qld) 1992* is limited to the historical (primarily non-indigenous cultural) environment. The Act creates a Heritage Council, which maintains a Heritage Register. The Council assesses development applications relating to registered places, although approval (or otherwise) can be delegated to local government authorities working within the *Integrated Planning Act (Qld) 1997*. While there is no provision for financial assistance to owners of heritage properties, Heritage Agreements may be entered into by the government and owners. The Cultural Heritage section of the Environment Protection Agency (EPA) provides professional support and services for the Council.

Protected natural areas are defined by reference to their management objectives, as set out in the *Nature Conservation Act (Qld) 1992*. National parks are required to permanently preserve an area's natural condition and its cultural and natural values, and any use must be nature-based and ecologically sustainable. In addition to 'generalist' national parks, individual parks may also be more specifically declared as 'scientific', 'Aboriginal land' or 'Torres Strait Islander land'. Conservation parks differ from national parks in that commercial uses such as fishing and grazing are allowed under certain conditions. The Parks and Forests section of the EPA administer most reserves.

Areas or features of 'anthropological, cultural, historic, prehistoric or societal significance' to Indigenous peoples are now covered by the *Aboriginal Cultural Heritage Act (Qld) 2003* and the *Torres Strait Islander Cultural Heritage Act (Qld) 2003*, administered somewhat incongruously by the Department of Natural Resources, Mines and Water. Anyone undertaking activities in an area must comply with the Aboriginal cultural heritage duty-of-care guidelines or enter into an agreement with the appropriate Aboriginal body or Traditional Owners.

South Australia (SA)

The *Heritage Places Act (SA) 1993*, as amended by the *Heritage (Heritage Directions) Amendment Act (SA) 2005*, is the legislation dealing with non-indigenous cultural (built) heritage, and is administered by the Heritage Branch (HB) of the Department for Environment and Heritage. The Act aims at the preservation, protection and enhancement of the physical, social and cultural heritage of the State, but it and the HB are not significantly concerned with either Aboriginal or natural heritage. The Branch administers the Heritage Register, provides administrative support for the Heritage Council, and provides advice to local councils and property owners. Larger areas, such as historic towns, streetscapes and natural areas, can be designated State Heritage Areas. Work on heritage properties requires approval from the local authority or Planning SA, depending on the scale at which the item is significant; the *Development Act (SA) 1993* provides guidance concerning locally significant items.

The key pieces of legislation concerning natural heritage are the *National Parks and Wildlife Act (SA) 1972* and the *Wilderness Protection Act (SA) 1992*. Administration of the latter is undertaken by the Wilderness Advisory Committee, while other conservation reserves are administered by National Parks and Wildlife SA. Aboriginal heritage is protected under the *Aboriginal Heritage Act 1988*, administered by the Department for Aboriginal Affairs and Reconciliation. The Minister is advised by a committee comprising Aborigines from around the State. Blanket protection is provided for all Aboriginal sites and objects that are 'significant according to Aboriginal tradition', though there is also a register of specific sites.

Tasmania (Tas)

Tasmanian heritage legislation and administration is in a state of flux at the beginning of the twenty-first century. Heritage is administered at the state level by various divisions of the Department of Tourism, Parks, Heritage and the Arts (DTPHA). Following a reorganisation in 2002–2003, Heritage Tasmania focuses primarily on historic heritage (i.e., built, non-indigenous cultural heritage), while a separate Aboriginal Heritage Office was created, and a Cultural Heritage Unit was established within the Tasmanian Parks and Wildlife Service (TPWS) to take care of historic sites under the *Parks and Reserves Management Act (Tas)* 2002. In addition, an extensive review of Tasmanian heritage legislation was undertaken, with the time for public comment concluding in October 2005. The major recommendations focused on providing greater clarity as to the relationship between heritage and planning legislation. A major review of legislation concerning Aboriginal heritage was underway in early 2006, with the purpose of providing 'effective recognition, assessment, protection and management of Aboriginal heritage, and the empowerment of the Tasmanian Aboriginal community' (TAHL 2006). It remained to be seen, in early 2006, what new legislation would result from these two review processes.

Historic places outside natural reserves are protected by the *Historic Cultural Heritage Act (Tas) 1995*, which established the Heritage Council and the Heritage Register; Heritage Tasmania supports the Council in implementing the Act. Natural reserves are administered and managed by the TPWS, operating under the *National Parks and Reserves Management Act (Tas) 2003*, while Aboriginal sites are protected under the *Aboriginal Relics Act (Tas) 1975*, administered by the Aboriginal Heritage Office of DTPHA. In 2006, the Office was establishing new policies and administrative systems to improve the standards of Aboriginal heritage management.

Victoria (Vic)

Legislative protection for heritage places is provided by the *Heritage Act (Vic) 1995*. In addition, local authorities have been increasingly active in recent years in the use of their planning powers to protect heritage places. The Act provides for the protection and preservation of buildings and other works, and of objects of architectural or historic significance, and for the Heritage Council and Heritage Register. It also deals specifically with archaeological sites and shipwrecks, but not with natural heritage. Heritage Victoria provides administrative and practical support for the Council, putting its policies and the provisions of the Act into operation. The *Planning Environment Act (Vic) 1987* makes it obligatory for planning schemes to be administered by local government so as to protect registered places. A new Victorian Heritage Strategy, dealing with cultural heritage, was due for release in 2006.

The *National Parks Act (Vic) 1975*, administered by Parks Victoria, allows for the gazettal of national parks and other types of reserve. Unlike those in most other states, Victorian parks are created by legislative, rather than executive or administrative, action. The objects of the Act include research and the identification of additional areas for reservation, the latter process assisted by the Land Conservation Council, which oversees the use of Crown land generally.

The considerable Federal Government powers in relation to Aboriginal heritage explain, in part, how the major legislation affecting conservation of Aboriginal heritage in Victoria is the *Aboriginal and Torres Strait Islander Heritage Protection Act (Cwlth) 1984*, as amended in 1987, administered within the State by Aboriginal

Affairs Victoria. This arrangement provides comprehensive protection for Indigenous heritage. In 2004, an Exposure Draft of a planned new Aboriginal Heritage Bill was released for comment. Among other things, if passed, this will: establish a Victorian Aboriginal Heritage Council; allow broader Aboriginal involvement; provide stronger and clearer legislative protection for Aboriginal cultural heritage; and link heritage legislation to the planning system. Public submissions closed in December 2005 and the final Bill was being drafted in early 2006.

Western Australia (WA)

The *Heritage of Western Australia Act (WA) 1990* establishes the Heritage Council, which, unlike its counterparts in other states, has its own Director and support staff rather than relying on a separate bureaucratic entity. The Council maintains the Register of Heritage Places. Heritage Agreements with owners, also available in a number of other states, play a more important role in WA. Proposals for development affecting a registered place must be referred to the Council for advice, though the final decision lies with the consent authority or the Minister. Advice from the Council must be followed if development takes place. Financial and other incentives are offered to owners of registered places to assist in conservation efforts, while conservation orders and restoration orders may be made.

National parks and nature reserves are vested in the Conservation Commission of WA, set up under the *Conservation and Land Management Act (WA) 1984*. While the Authority has the responsibility to prepare plans of management, the Department of Conservation and Land Management (CALM) has managed the reserves (now Department of Environment and Conservation, see Chapter 6). The Department of Indigenous Affairs administers the *Aboriginal Heritage Act (WA) 1972*, which provides for all known and unknown Aboriginal and archaeological sites being the property of the State. The Department is responsible for the care and protection of such places, while it is an offence to excavate, destroy, damage, conceal or alter any site, or to deal with them in a way that is contrary to Aboriginal custom, unless specific consent has been given. It also maintains the Register of Aboriginal Sites.

Voluntary and community organisations

The National Trust of Australia, with branches in each state and territory, is the premier voluntary organisation concerned with cultural heritage and the built environment, although it also plays an important role in natural heritage issues. The first state branch was established in NSW in the mid-1940s, others following later, so they predate government heritage bodies by up to 30 years. State Trust branches are now usually represented on government heritage bodies. All of the branches, except that in South Australia, maintain registers of their own, while all lobby hard for conservation of heritage items in general, and of those under threat in particular. Branches have also acquired a small number of properties, and have helped restore others, to help achieve these objectives. Not all Trust properties are open to the public on a regular basis, as the Trust needs to gain income through leases or to otherwise work in conjunction with other owners or occupiers. Branches also typically have a committee system, which investigates heritage issues. Trust listing in itself has no legal force, but does carry strong moral persuasion powers and is taken into account in any planning application. At the national level, the Australian Council of National Trusts is represented on federal bodies, lobbies strongly on federal issues, and was a key player in deliberations leading up to the drafting of new Commonwealth heritage legislation. One way or another, a large number of dedicated volunteers give time and effort to help achieve the Trust's objectives.

There are also many other bodies that play a key role in heritage conservation through the research, notification, lobbying and persuasive pressure they bring to bear. A few of the more obvious are: the Royal Australian Institute of Architects, which, in NSW at least, has an influential list of important 20th-century buildings (often neglected by both the government bodies and the National Trust); the Institute of Engineers (industrial and engineering heritage); National Parks Association (natural areas, and, in NSW, support for NPWS activities); Wilderness Society; local historical and conservation associations (see Chapter 11); local progress associations and citizens' committees (but some are very pro-development); ethnic communities' organisations (recently enlisted by the NSW Heritage Office to 'de-Anglicise' the state's listings); Aboriginal Lands Councils and similar bodies representing Indigenous peoples; specialist groups interested in, for example, railways, theatres and theatre organs, and gardens; and religious organisations and local congregations.

Conclusion

Heritage became a public concern in Australia only quite recently, and still ranks well below development, growth, progress, and the economy in national priorities. Due to the political realities of the Australian federation, legislative and bureaucratic frameworks are complex and, because so much responsibility rests with state, territory and local government, vary from place to place. Recent federal legislation has clarified the relationship between the Federal Government, on the one hand, and the state and territory governments on the other, while the role of local government remains dependent on state and territory legislation. Another complicating factor involves the ways in which the three key types of heritage – natural, Indigenous, and non-indigenous cultural – are often separated administratively, even though distinctions between them are typically far less clear in the field.

Despite the recency of this interest in heritage, Australia generally has a robust system of identification, conservation, and management of heritage. However, as already stated, heritage does often play second fiddle to other concerns, and there is a great deal of scope for political interference and over-ruling of decisions by the heritage agencies, especially on the interface between heritage and planning.

References

Aplin, G. (2002) *Heritage: Identification, Conservation, and Management* (Melbourne: Oxford University Press).

Australian Heritage Commission (1981) *The Heritage of Australia: The Illustrated Register of the National Estate* (Melbourne: Macmillan).

Australian Heritage Commission (1997) *Annual Report 1996–97* (Canberra: Australian Heritage Commission).

NSW Heritage Office (2004) *Heritage and Sustainability* (Discussion Paper) (Sydney: Heritage Office of NSW).

Websites

Much of the material for this chapter was obtained from government websites, the most important of which are listed below. All of these sites were accessed in late March or early April 2006.

Australia (federal)

Australian Heritage Directory: http://www.heritage.gov.au/
Environment Australia heritage site: http://www.deh.gov.au/heritage/index.html
Australian Heritage Council: http://www.ahc.gov.au/
National Heritage List: http://www.deh.gov.au/heritage/national/index.html
Biodiversity pages, DEH: http://www.deh.gov.au/biodiversity/index.html
Parks and Reserves pages, DEH: http://www.deh.gov.au/parks/index.html
Australia's World Heritage: http://www.deh.gov.au/heritage/worldheritage/
New legislation, EPBC Homepage: http://www.deh.gov.au/epbc/index.html

Australian Capital Territory

Environment ACT heritage pages: http://www.environment.act.gov.au/heritage
Parks: http://www.environment.act.gov.au/bushparksandreserves
ACT Heritage: http://www.environment.act.gov.au/heritage/actheritage
ACT Heritage Council: http://www.environment.act.gov.au/heritage/heritagecouncil

New South Wales

Department of Planning: http://www.planning.nsw.gov.au/index.html
NSW Heritage Office and Heritage Council of NSW: http://www.heritage.nsw.gov.au
Heritage listings: http://www.heritage.nsw.gov.au/07_subnav_04.cfm
Historic Houses Trust: http://www.hht.net.au/
National Parks and Wildlife Service: http://www.nationalparks.nsw.gov.au/

Northern Territory

Department of Natural Resources, Environment and the Arts heritage pages:http://www.nt.gov.au/nreta/heritage/index.html
NT Heritage Register: http://www.nt.gov.au/nreta/heritage/ntregister/index.html
Aboriginal Areas Protection Authority: http://www.nt.gov.au/aapa/

Parks: http://www.nt.gov.au/nreta/parks//

Queensland

Cultural heritage: http://www.epa.qld.gov.au/cultural_heritage
Registers and inventories: http://www.epa.qld.gov.au/cultural_heritage/registers_
    and_inventories
Queensland Heritage Council:
http://www.epa.qld.gov.au/cultural_heritage/registers_and_inventories/queensland_
    heritage_council
Indigenous Heritage: http://www.epa.qld.gov.au/cultural_heritage/indigenous_
    heritage
National Parks: http://www.epa.qld.gov.au/parks_and_forests

South Australia

Heritage Branch: http://www.environment.sa.gov.au/heritage/
SA Heritage Council: http://www.environment.sa.gov.au/heritage/authority.html
State Heritage Register: http://www.environment.sa.gov.au/heritage/register.html
National Parks and Wildlife SA: http://www.parks.sa.gov.au/
Department for Aboriginal Affairs and Reconciliation: http://www.daare.sa.gov.au

Tasmania

Tasmanian Parks and Wildlife Service: http://www.parks.tas.gov.au/
Dept of Tourism, Parks, Heritage and the Arts: http://www.dtpha.tas.gov.au/
Aboriginal Heritage Office (of DTPHA): http://www.dtpha.tas.gov.au/divisions_
    aho.html
Heritage Tasmania: http://www.dtpha.tas.gov.au/divisions_ht.html

Victoria

Department of Sustainability and Environment heritage pages: http://www.dse.vic.
    gov.au/dse/dsenher.nsf
Heritage Council of Victoria: http://www.heritage.vic.gov.au/
Victorian Heritage Register: http://www.doi.vic.gov.au/doi/hvolr.nsf
Parks Victoria: http://www.parkweb.vic.gov.au/
Aboriginal Affairs Victoria:http://www1.dvc.vic.gov.au/aav/

Western Australia

Department of Environment and Conservation: http://www.calm.wa.gov.au/
Heritage Council of WA: http://www.heritage.wa.gov.au/
Heritage Register: http://register.heritage.wa.gov.au/quicksearch.html
Department of Indigenous Affairs: http://www.dia.wa.gov.au/
National Parks: http://www.calm.wa.gov.au/national_parks/

Australian voluntary organisations

Australian Council of National Trusts: http://www.nationaltrust.org.au/
National Trust of Australia (ACT): http://www.act.nationaltrust.org.au/
National Trust of Australia (NT): access through national (ACNT) site above.
National Trust of Australia (NSW): http://www.nsw.nationaltrust.org.au/
National Trust of Australia (Qld): http://www.nationaltrustqld.org/
National Trust of Australia (Tas):http://www.discovertasmania.com.au/home/index.
    cfm?SiteID=223/
National Trust of Australia (Vic): http://www.nattrust.com.au/
National Trust of Australia (WA): http://www.ntwa.com.au/
Nature Conservation Council of NSW: http://www.nccnsw.org.au/
National Parks Association of NSW: http://www.npansw.org.au/

Chapter 3

Australia and World Heritage

Graeme Aplin

The concept of 'World Heritage'

Some heritage items are seen as significant at the global scale – as having significance for the world's population – although individual perceptions of significance obviously differ. The World Heritage Committee (WHC) and its advisers deem World Heritage properties significant globally for one or more of an accepted list of reasons (Tables 3.1 and 3.2), acting within structures established under the *Convention Concerning the Protection of the World Cultural and Natural Heritage*. The World Heritage Convention, as it is known, is administered through the United Nations Educational, Scientific and Cultural Organization (UNESCO), was adopted in 1972, and came into force in December 1975. By the end of 2005, there were 181 State Parties and 812 properties inscribed on the World Heritage List (WHL), made up of 628 cultural sites, 160 natural sites, and 24 mixed sites, distributed over 137 State Parties. The Convention's objective is 'to establish an effective system of collective protection of the cultural and natural heritage of outstanding universal value, organized on a permanent basis and in accordance with modern scientific methods'. The primary responsibilities for identifying and protecting heritage, including World Heritage, remain with the State in which it is located. Once a property is inscribed on the WHL, however, the WHC has much greater power to influence the 'host' government.

A General Assembly of State Parties meets every two years and elects the WHC, comprising one representative from each of 21 State Parties chosen to represent the major regions and cultures of the world. The Committee meets once a year, with on-going administrative functions carried out by the smaller seven-person World Heritage Bureau (WHB), which meets formally at least twice each year. A small number of staff housed at the UNESCO World Heritage Centre in Paris supports the work of the Committee and Bureau.

Identifying, assessing and managing World Heritage

The Convention, in Articles 1 and 2, defines cultural and natural heritage, respectively, as indicated in Table 3.1, emphasising the 'universal value' that must be present. The WHC has approved Operational Guidelines (modified over time) that include more precisely worded selection criteria for the assessment of nominations; Table 3.2 gives them as they were before combination into one list in 2005. Cultural properties must meet one or more of criteria C(i) to C(vi), while natural properties must meet one or more of N(i) to N(iv). Additional conditions must also be met: natural heritage properties must have integrity; cultural sites must be authentic; and all sites must have adequate legal or institutional protection and plans of management. In other words, properties must not only possess global significance, but there must be the means to maintain it.

Table 3.1   Definitions of cultural and natural heritage, from Articles 1 and 2 of the World Heritage Convention (emphasis added)

Cultural Heritage (Article 1)

monuments: architectural works, works of monumental sculpture and painting, elements or structures of an archaeological nature, inscriptions, cave dwellings and combinations of features, *which are of outstanding universal value* from the point of view of history, art or science;

groups of buildings: groups of separate or connected buildings which, because of their architecture, their homogeneity or their place in the landscape, *are of outstanding universal value* from the point of view of history, art or science;

sites: works of man or the combined works of nature and man, and areas including archaeological sites *which are of outstanding universal value* from the historical, aesthetic, ethnological or anthropological point of view.

Natural Heritage (Article 2)

natural features consisting of physical and biological formations or groups of such formations, *which are of outstanding universal value* from the aesthetic or scientific point of view;

geological and physiographical formations and precisely delineated areas which constitute the habitat of threatened species of animals and plants *of outstanding universal value* from the point of view of science or conservation;

natural sites or precisely delineated natural areas *of outstanding universal value* from the point of view of science, conservation or natural beauty'.

Source: Convention Concerning the Protection of the World Cultural and Natural Heritage.

State Parties can nominate properties for inclusion if they meet one or more of the criteria – natural, cultural, or a combination. Properties can also be renominated to increase their area or meet additional criteria. The WHB assesses each nomination based on an evaluation by one or both of the two key international advisory bodies: the International Council on Monuments and Sites (ICOMOS), and the World Conservation Union (IUCN). After evaluation and assessment, the WHB makes recommendations regarding inscription to the WHC, which assesses and votes on each proposal. The Bureau and Committee also maintain an on-going interest in the state of conservation of all properties on the WHL.

National governments need to pass legislation giving effect to their ratification of the Convention. Once a property is inscribed, ownership and control remain essentially as they were, rather than passing to the WHC. However, whatever a property's ownership, the State Party is responsible for ensuring that it is managed in a way that maintains its World Heritage values. There are also more general

Table 3.2   Criteria for inscription on the World Heritage List (before 2005 revision)

Cultural properties should meet one or more of the following criteria:

C(i) represent a masterpiece of human creative genius; or

C(ii) exhibit an important interchange of human values, over a span of time or within a cultural area of the world, on developments in architecture or technology, monumental arts, town planning or landscape design; or

C(iii) bear a unique or at least exceptional testimony to a cultural tradition or to a civilisation which is living or which has disappeared; or

C(iv) be an outstanding example of a type of building or architectural or technological ensemble or landscape which illustrates a significant stage or significant stages in human history; or

C(v) be an outstanding example of a traditional human settlement or land use which is representative of a culture or cultures, especially when it has become vulnerable under the impact of irreversible change; or

C(vi) be directly or tangibly associated with events or living traditions, with ideas, with beliefs, or with artistic and literary works of outstanding universal significance (a criterion used only in exceptional circumstances, or together with other criteria).

Natural properties should meet one or more the following criteria:

N(i) be outstanding examples representing major stages of the earth's history, including the record of life, significant ongoing geological processes in the development of landforms, or significant geomorphic or physiographic features; or

N(ii) be outstanding examples representing significant ongoing ecological and biological processes in the evolution and development of terrestrial, fresh water, coastal and marine ecosystems and communities of plants and animals; or

N(iii) contain superlative natural phenomena or areas of exceptional natural beauty and aesthetic importance; or

N(iv) contain the most important and significant natural habitats for *in situ* conservation of biological diversity, including those containing threatened species of outstanding value from the point of view of science or conservation.

Source: Operational Guidelines for the Implementation of the World Heritage Convention, 2004.

obligations on each State Party to promote the cause of heritage conservation in that country and, through international co-operation, in other countries.

The WHC also prepares the List of World Heritage in Danger (LWHD), which contains properties thought to be in 'serious and specific danger' of losing their World Heritage values. In late 2005, the List contained 34 properties, listed for four main reasons: natural environmental threats; human threats relating to development and use, including visitor pressure; war and civil unrest; and lack of adequate management and conservation plans and/or adequate legislative and bureaucratic support. A number of properties on the LWHD have been subsequently removed as threats have ceased, or at least have been mitigated.

Australia and the World Heritage Convention

Australia ratified the Convention in 1974, one of the first nations to do so. Since then it has played an active role in the deliberations of the WHB and WHC, and in many other ways. Australia was a member of the WHC from its initiation in 1977 until 1988, serving two six-year terms, and for another term in 1995–2000. Australia acted as Chairperson in 1981–82 and 2000, and provided one of the Vice-Chairpersons in 1980, 1983–84, 1988, and 1995–96. Australia has had a least one representative at every WHC meeting, and has played a prominent role on various committees and working groups examining particular issues.

The management objectives set by the Federal Government to meet its obligations in relation to World Heritage properties are:

1. to protect, conserve and present the World Heritage values of the property;
2. to integrate the protection of the area into a comprehensive management program;
3. to give the property a function in the life of the Australian community;
4. to strengthen appreciation and respect of the property's World Heritage values, particularly through educational and information programs;
5. to keep the community broadly informed about the condition of the World Heritage values of the property; and
6. to take appropriate scientific, technical, legal, administrative and financial measures necessary for the achieving of the foregoing objectives.

The legislative backing for these objectives was the *World Heritage Properties Conservation Act (Cwlth) 1983* until 1999, when it was replaced by the relevant sections of the *Environment Protection and Biodiversity Conservation Act (Cwlth) 1999* (EPBCA, see Chapter 2). While the Federal Government, as State Party, is ultimately responsible for maintaining World Heritage values, individual properties are managed in various ways, as described below.

As part of the Asia–Oceania Periodic Reporting exercise in 2002–03, the Australian Government, through DEH, produced reports on the state of conservation of all its World Heritage sites inscribed prior to 1995 (Australia 2003). Some brief comments on individual properties are included below. The report also lists several general, high-priority actions, including: improving management plans; improving training opportunities for on-site managers; enhancing opportunities for the involvement of Indigenous people; enhancing the Asia-Pacific Focal Point, especially in the Pacific (see below); and achieving greater investment in collaborative research and improved funding arrangements. The Asia-Pacific Focal Point, established in 1999 and hosted by Australia, is an information-sharing scheme to help Asian and Pacific countries (including ones not yet State Parties) improve management of World Heritage and other heritage properties, prepare nominations for inscription on the WHL, and

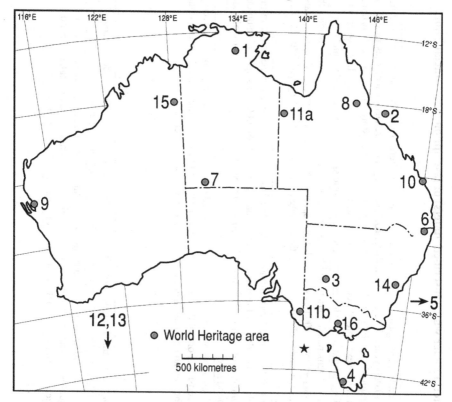

Figure 3.1   Australian World Heritage properties

Note: Only the general locations of larger properties are indicated, not their extent, while properties 6 and 8 each have many component parts spread over a wide area. Key: 1. Kakadu National Park; 2. Great Barrier Reef; 3. Willandra Lakes Region; 4. Tasmanian Wilderness; 5. Lord Howe Island Group; 6. Central Eastern Australian Rainforest Reserves; 7. Uluru–Kata Tjuta National Park; 8. Wet Tropics of Queensland; 9. Shark Bay, Western Australia; 10. Fraser Island; 11. Australian Fossil Mammal Sites (Riversleigh (11a)/Naracoorte (11b); 12. Heard and McDonald Islands; 13. Macquarie Island; 14. The Greater Blue Mountains Area; 15. Purnululu National Park; 16. Royal Exhibition Building and Carlton Gardens.

generally increase their knowledge and understanding of, and engagement with, the World Heritage process. In 2004, the WHC specifically congratulated Australia and Indonesia on reaching an agreement to co-operate in the management of the Wet Tropics of Queensland and Lorentz National Park, essentially a way for Australia to provide assistance to Indonesia.

Australia's World Heritage properties

Table 3.3 lists Australia's sixteen World Heritage properties, with their date(s) of inscription, the criteria under which they were judged significant (see Table 3.2), and a brief statement of the reasons for inscription. These properties are mapped

Table 3.3 Australia's World Heritage properties, with criteria under which inscribed, date(s) of inscription, and 'official' short description

Property	Criteria	Date(s)	Description
Kakadu National Park	N(ii), N(iii), N(iv); C(i), C(vi)	1981, 1987, 1992	This unique archaeological and ethnological reserve, located in the Northern Territory, has been inhabited continuously for more than 40,000 years. The cave paintings, rock carvings and archaeological sites record the skills and way of life of the region's inhabitants, from the hunter-gatherers of prehistoric times to the Aboriginal people still living there. It is a unique example of a complex of ecosystems, including those of tidal flats, floodplains, lowlands and plateau, providing habitat for a wide range of rare or endemic species of plants and animals.
Great Barrier Reef	N(i), N(ii), N(iii), N(iv)	1981	The Great Barrier Reef is a site of remarkable variety and beauty on the north-east coast of Australia. It contains the world's largest collection of coral reefs, with 400 types of coral, 1,500 species of fish, and 4,000 types of mollusc. It also holds great scientific interest, as the habitat of species such as the dugong ('sea cow') and the large green turtle, which are threatened with extinction.
Willandra Lakes Region	N(i); C(iii)	1981	The fossil remains of a series of lakes and sand formations that date from the Pleistocene can be found in this region, together with archaeological evidence of human occupation dating from 45–60,000 years ago. It is a unique landmark in the study of human evolution on the Australian continent. Several well-preserved fossils of giant marsupials have also been found here.
Tasmanian Wilderness	N(i), N(ii), N(iii), N(iv); C(iii), C(iv), C(vi)	1982, 1989	In a region that has been subjected to severe glaciation, these parks and reserves, with their steep gorges, covering an area of over 1 million ha, constitute one of the last expanses of temperate rainforest in the world. Remains found in limestone caves attest to the human occupation of the area for more than 20,000 years.
Lord Howe Island Group	N(iii), N(iv)	1982	A remarkable example of isolated oceanic islands, born of volcanic activity more than 2,000 m under the sea, these islands boast a spectacular topography and are home to numerous endemic species, especially birds.
Central Eastern Australian Rainforest Reserves	N(i), N(ii), N(iv)	1986, 1994	This site, comprising several protected areas, is situated predominantly along the Great Escarpment on Australia's east coast. The outstanding geological features displayed around shield volcanic craters and the high number of rare and threatened rainforest species is of international significance for science and conservation.

Site	Criteria	Year	Description
Uluṟu–Kata Tjuṯa National Park	N(ii), N(iii); C(v), C(vi)	1987, 1994	The park, formerly called Uluru (Ayers Rock–Mount Olga) National Park, features spectacular geological formations that dominate the vast red sandy plain of central Australia. Uluru, an immense monolith, and Kata Tjuṯa, the rock domes located west of Uluru, form part of the traditional belief system of one of the oldest human societies in the world. The traditional owners of Uluṟu–Kata Tjuṯa are the Anangu Aboriginal people.
Wet Tropics of Queensland	N(i), N(ii), N(iii), N(iv)	1988	This area, which stretches along the north-east coast of Australia for some 450 km, is made up largely of tropical rainforests. This biotope offers a particularly extensive and varied array of plants, as well as marsupials and singing birds, along with other rare and endangered animals and plant species.
Shark Bay, Western Australia	N(i), N(ii), N(iii), N(iv)	1991	At the most westerly point of the Australian continent, Shark Bay, with its islands and the land surrounding it, has three exceptional natural features: its vast sea-grass beds, which are the largest (4,800 square kilometres) and richest in the world; its dugong ('sea cow') population; and its stromatolites (colonies of algae which form hard, dome-shaped deposits and are among the oldest forms of life on earth). Shark Bay is also home to five species of endangered mammals.
Fraser Island	N(ii), N(iii)	1992	Fraser Island lies just off the east coast of Australia. At 122 km long, it is the largest sand island in the world. Majestic remnants of tall rainforest growing on sand and half the world's perched freshwater dune lakes are found inland from the beach. The combination of shifting sand dunes, tropical forests and lakes make it an exceptional site.
Australian Fossil Mammal Sites (Riversleigh/Naracoorte)	N(i), N(ii)	1994	Riversleigh and Naracoorte, in the north and south respectively of eastern Australia, are among the world's ten greatest fossil sites. They are a superb illustration of the stages of evolution of Australia's unique fauna.
Heard and McDonald Islands	N(i), N(ii)	1997	Heard Island and McDonald Islands are located in the Southern Ocean, approximately 1,700 km from the Antarctic continent and 4,100 km south-west of Perth. As the only volcanically active subantarctic islands they 'open a window into the earth', thus providing the opportunity to observe ongoing geomorphic processes and glacial dynamics. The distinctive conservation value of Heard and McDonald – one of the world's rare pristine island ecosystems – lies in the complete absence of alien plants and animals, as well as of human impact.

Table 3.3 continued

Macquarie Island	N(i), N(iii)	1997	Macquarie Island (34 km long x 5 km wide) is an oceanic island in the Southern Ocean, lying 1,500 km south-east of Tasmania and approximately half way between Australia and the Antarctic continent. The island is the exposed crest of the undersea Macquarie Ridge, raised to its present position where the Indo-Australian tectonic plate meets the Pacific plate. It is a site of major geoconservation significance, being the only place on earth where rocks from the earth's mantle (6 km below the ocean floor) are being actively exposed above sea-level. These unique exposures include excellent examples of pillow basalts and other extrusive rocks.
The Greater Blue Mountains Area	N(ii), N(iv)	2000	The Greater Blue Mountains Area consists of 1.03 million ha of sandstone plateaux, escarpments and gorges dominated by temperate eucalypt forest. The site, comprised of eight protected areas, is noted for its representation of the evolutionary adaptation and diversification of the eucalypts in post-Gondwana isolation on the Australian continent. Ninety-one eucalypt taxa occur within the Greater Blue Mountains Area, which is also outstanding for its exceptional expression of the structural and ecological diversity of the eucalypts associated with its wide range of habitats. The site provides significant representation of Australia's biodiversity with ten per cent of the vascular flora as well as significant numbers of rare or threatened species, including endemic and evolutionary relict species, such as the Wollemi pine, which have persisted in highly-restricted microsites.
Purnululu National Park	N(i), N(iii)	2003	The 239,723 ha Purnululu National Park is located in the State of Western Australia. It contains the deeply dissected Bungle Bungle Range composed of Devonian-age quartz sandstone eroded over a period of 20 million years into a series of beehive-shaped towers or cones, whose steeply sloping surfaces are distinctly marked by regular horizontal bands of dark-grey cyanobacterial crust (single-celled photosynthetic organisms). These outstanding examples of cone karst owe their existence and uniqueness to several interacting geological, biological, erosional and climatic phenomena.
Royal Exhibition Building and Carlton Gardens	C(ii)	2004	The Royal Exhibition Building and its surrounding Carlton Gardens were designed for the great international exhibitions of 1880 and 1888 in Melbourne. The building and grounds were designed by Joseph Reed. The building is constructed of brick and timber, steel and slate. It combines elements from the Byzantine, Romanesque, Lombardic and Italian Renaissance styles. The property is typical of the international exhibition movement that saw over 50 exhibitions staged between 1851 and 1915 in venues including Paris, New York, Vienna, Calcutta, Kingston (Jamaica) and Santiago (Chile). All shared a common theme and aims: to chart material and moral progress through displays of industry from all nations.

Source: World Heritage Centre website at http://whc.unesco.org/ accessed during March–April 2006.

in Figure 3.1. It is worth noting that four Australian properties – Kakadu, Uluru-Kata Tjuta, South-west Tasmania, and Willandra Lakes – have been inscribed for both natural and cultural heritage reasons. This 'dual inscription' is quite rare on a world scale and reflects the intimate connection between Australia's Indigenous peoples and their country. Furthermore, Uluru-Kata Tjuta has been inscribed as a cultural landscape, a categorisation that further emphasises that relationship. It is important to note that no sites were inscribed for non-indigenous cultural heritage significance, either alone or in conjunction with natural heritage significance, until the Royal Exhibition Building in 2005. Natural heritage sites always have been more dominant in Australia than in almost any other country. Some brief comments on each site follow more general discussion.

There has been much political controversy over World Heritage in Australia, at times overshadowing the undoubted values of the properties. Much of this has involved disagreement between the Federal Government and state governments over nominations, with states claiming inadequate consultation. Following extensive discussions, the 1992 Intergovernmental Agreement on the Environment required the Commonwealth to consult the states on World Heritage proposals, a step now enshrined in the *Environmental Protection and Biodiversity Conservation Act (Cwlth) 1999*. A second bone of contention has been the possible limitations on land uses such as forestry and mining imposed to meet Australia's obligations under the Convention. Furthermore, conflicts of interest arise because listing does not change ownership, and much of the day-to-day management is necessarily carried out by state agencies.

Kakadu National Park

Kakadu was inscribed in three stages, mirroring the evolution of the National Park. It is significant as an outstanding example of on-going geological and biological processes and of human interaction with the natural environment; as an example of superlative natural phenomena; as a site important for conservation of habitats and rare and endangered species; as a representation of unique artistic achievement; and as an area directly associated with ideas or beliefs of outstanding universal significance. The last two points relate to the occupance and culture of the Traditional Owners. The Park contains a wide variety of landscapes, including floodplains, hills and plateaux (or 'stone country'), prominent escarpments, and magnificent waterfalls. There are important habitats for a diverse range of flora and fauna, including many rare, endangered or endemic species. Aboriginal peoples have lived in the region for 40–60,000 years, and there are a large number of sacred sites and extensive art sites. The Park is leased from its Traditional Owners, who have a majority on the Board of Management, and is managed by Parks Australia. A Draft Management Plan 2006 has been open for public consultation and will eventually replace the 1999 Plan of Management.

Kakadu is unusual in having mining leases – areas of actual or potential uranium mining – within it. While not legally part of the World Heritage property, these enclaves are surrounded by it. In the late 1990s, proposals to allow a new mine to begin production were accepted by the Federal Government. Conservation groups and Traditional Owners moved to have Kakadu placed on the LWHD on the basis of potential environmental and cultural impacts and a significant risk of diminution of World Heritage values. The WHC did not place Kakadu on the LWHD, but came extremely close to doing so in 1999, voicing major concerns about Australia's handling of the issue. The 1999 Extraordinary Meeting of the WHC followed the sending of a mission to Kakadu and its report, which included a number of recommendations (UNESCO 1998). The Australian Government responded vigorously through its own report (Environment Australia 1999), and by sending an unusually large delegation to the meeting. The Traditional Owners and conservation groups also sent representatives and lobbied hard. The Federal Government did take a number of actions in response to the Mission's recommendations, but also expressed concern and obvious displeasure at some, arguing, among other things, that the Mission did not adequately defer to the role of the State Party in management decisions. Many groups remained dissatisfied and continued calls for an 'in-danger' listing and a ban on mining expansion. Inscription on the LWHD was again avoided in 2000, and the issue has since largely disappeared from the WHC meeting agendas. A much more detailed analysis of this debate is contained in Aplin (2004). In 2006, however, the general issue of uranium mining and exports was back on the Australian political agenda; whether or not this has direct implications for Kakadu remains to be seen, even though the Australian Government and the mining company have promised repeatedly that mining at Jabiru will only proceed if the Traditional Owners agree.

The WHC on a number of occasions has expressed concern about leaks of contaminated water from the Ranger mine site, even though Australia repeatedly has insisted that there have been no adverse impacts on the World Heritage property. Another cause for concern in 2002–03 was the perceived need to improve consultation with Traditional Owners over cultural heritage management. In 2005, the WHC heard of cane toad problems, and commended Australia for its attempts to improve tourism management in Kakadu. The Australian Government is still mentioning the possibility of renomination as a cultural landscape, possibly covering an extended area.

Great Barrier Reef

The Great Barrier Reef is one of very few properties inscribed for all four natural criteria. It illustrates important geological and biological processes, is an example of superlative natural phenomena, and is crucial for conservation of biodiversity, particularly given the wide variety of habitats it contains. The property includes some 2,800 individual reefs extending over 2,000 km and covering an area of 35 million ha (larger than Italy); it is by far the world's largest coral reef area, and the largest World Heritage Area. Almost all of the property is in the Great Barrier Reef Marine Park, declared in 1975 and managed by a Commonwealth agency, the Great Barrier Reef Marine Park Authority (GBRMPA), but with the Queensland Parks and Wildlife Service (QPWS) providing day-to-day management. Both Commonwealth and State legislation applies. Management is very complex, given the potential conflicts between uses that include conservation, tourism, and commercial fisheries, and a system of zoning is central to this management. Climate change is a potentially serious threat as warmer water in the shallow seas of the Reef may lead to extensive coral bleaching and a serious loss of biodiversity; associated changes in sea level will also impact adversely. Outflows of agricultural chemicals and silt from mainland farming districts constitute another serious problem, while infrequent but potentially disastrous shipping accidents are of great concern. The WHC has expressed its concern over a number of issues over the years, including the building of the Cape Tribulation road through the Daintree rainforest, large-scale resort development proposals, acid sulphate soils in coastal areas, possible development of oil-shale deposits on land and oil and gas exploration at sea, fisheries exploitation, and the need for a comprehensive planning regime. The Australian and Queensland governments, largely through GBRMPA, have responded positively to these concerns. Pollution from adjacent mainland areas remains perhaps the greatest problem, even though arising outside the property; a Water Quality Action Plan was prepared in 2001, but its success depends on the cooperation of many individuals and organisations. A petition was presented to the 2005 WHC meeting to have this property added to the LWHD, but no discussion ensued.

Willandra Lakes Region

The criteria justifying inscription involved the property representing the major stages in the earth's evolutionary history and significant on-going geological processes, and bearing an exceptional testimony to a past civilisation. It contains a system of Pleistocene lakes (now saline plains) and fringing dunes and lunettes, including the famous Walls of China. It is now accepted that Aborigines have lived on the lakeshores for more than 60,000 years, with DNA being recovered from a skeleton from that time. In 1968, excavations uncovered the cremation of a woman from 50–40,000 years ago, the earliest known cremation in the world, and burials thought to date from 45,000 years ago have also been found. There is abundant other evidence of human occupation throughout this long period, with radiocarbon dates verifying its antiquity. Remains of large marsupials have also been found. The WHC was initially concerned at the lack of a management plan, but that was rectified, though not until 1996. The property is primarily managed by the NSW National Parks and Wildlife Service (NPWS), even though not all of it is within Mungo National Park. A series of consultative and advisory committees involve Traditional Owners and pastoralists, among others. In 1995, the WHC accepted an Australian proposal for a revised boundary that reduced the property's size by 30 per cent, but better defined the area containing World Heritage values and facilitated better management.

Tasmanian Wilderness

This property was inscribed under all four natural criteria and three of the cultural ones. In terms of cultural significance, it bears exceptional testimony to a civilisation or cultural tradition, is a landscape that illustrates significant stages in human history, and is directly and tangibly associated with living traditions of outstanding universal significance. The property covers 1.38 million ha, or approximately 20 per cent of Tasmania. Rocks from the Precambrian and all subsequent periods are represented and there are extensive limestone cave systems. The area contains a wide variety of vegetation, including both Antarctic and Australian elements, and is recognised by IUCN as an International Centre for Plant Diversity. There is also an unusually high proportion of endemic faunal species, as well as relict groups of ancient lineage. Archaeological surveys have revealed an exceptionally rich collection of Aboriginal sites, the best known of which is Kutikina Cave. Most of the property is included in national parks and other reserves managed by the Tasmanian Parks and Wildlife Service.

The Tasmanian Wilderness World Heritage Area arose from a heated conflict over the Gordon-below-Franklin Dam proposal. The Tasmanian version of developmentalism (see Chapter 2) was based on 'hydroindustrialisation', or attracting industry through very cheap hydro-electric power. In 1979 a report recommended the Gordon-below-Franklin Dam and conservationists were determined not to lose another battle, as they had done previously over the Lake Pedder dam. In 1982, the Western Tasmanian Wilderness National Parks were inscribed on the WHL, after the Tasmanian Government had recommended nomination to the Commonwealth. The new status gave the Commonwealth power to protect the World Heritage values of the property. After surviving a Tasmanian legal challenge, the *World Heritage Properties Conservation Act (Cwlth) 1983* was passed; one effect was to effectively halt development of the Gordon-below-Franklin dam. When the property was initially inscribed in 1982, the WHC expressed concern about the possible impact of dam construction, and suggested that Australia ask it to place the property on the LWHD until the dam issue was resolved. Even though the new Commonwealth legislation prevented the dam going ahead, and Australia received the commendation of the Bureau, the LWHD suggestion was repeated as the effectiveness of the legislation still had to be tested. The property was extended in 1989. In the mid-1990s, the WHC expressed concern over logging and road building in areas adjacent to the property, but by the late 1990s, the Tasmanian Regional Forestry Agreement was seen as a positive step that might result in further extensions to the property. Various proposals for other developments that could impact on the property were also discussed in the 1999–2001 period, but the 2003 Australian report is very positive and no major concerns are expressed.

Lord Howe Island Group

Inscription was justified on the basis of the islands being an example of superlative natural phenomena, and due to their importance for biodiversity conservation. There are 105 endemic plant species and many of the bird species are rare or endangered, with the Lord Howe Island woodhen being one of the world's rarest birds. Introduced rats, feral cats, and owls are major problems. Two towering volcanic mountains are central to the spectacular nature of the landscape, while the surrounding waters have an unusual mixture of temperate and tropical organisms, and a number of smaller islands are also included. A Marine Park was declared by NSW in 1999, while a Commonwealth Marine Park surrounds the coastal State one. The need for more secure funding arrangements is the main concern expressed in the Australian 2003 report.

Central Eastern Australian Rainforest Reserves

This property, inscribed on the basis of geological and biological processes and the potential for biodiversity conservation, has about 50 constituent parts in north-eastern NSW and south-eastern Queensland, from the Barrington Tops just north of the Hunter Valley to the Lamington National Park. The 1994 extension also involved a change of name from the earlier Australian East Coast Subtropical and Temperate Rainforest Parks property, which had been limited to NSW. The WHB had recommended the extension in 1986 when the original inscription was approved, but the delay was due to strained relations between the Commonwealth and Queensland over World Heritage matters. The component parts are protected in a large number of national parks and other reserves managed by agencies of the two states. The property contains the world's most extensive subtropical rainforest and large portions of the world's Antarctic beech cool temperate rainforest, including much of the remaining forest of these types in eastern Australia. Many rare and threatened species of flora and fauna live in the reserves. The 2003 Australian report emphasises the importance of ongoing monitoring, especially in relation to potential loss of biodiversity.

Uluru–Kata Tjuta National Park

Uluru–Kata Tjuta was originally inscribed in 1987 on the grounds of natural criteria alone, but was then renominated and inscribed in 1994 on additional cultural criteria; it was also accepted under the relatively new category of cultural landscapes. In fact, when first inscribed in 1987, the WHC commended the Australian Government for its innovative management approach that blended natural and cultural elements of the park. Originally inscribed for its geological importance and natural beauty, it was subsequently also inscribed because it is one of the most ancient managed landscapes in the world, an outstanding illustration of human adaptation over many millennia to a hostile environment, and an integral part of the traditional belief system of one of the oldest human societies in the world. The National Park is leased back from the Traditional Owners, who have a majority on the Board of Management, and is managed by Parks Australia. Visitor pressure has been identified as one concern. In the 2003 report, the Australian Government also identified the need for close monitoring of cultural values and an increased emphasis on them in interpretation for visitors. An issue of practical concern is the Traditional Owners' desire to have visitors not climb Uluru, one of their most sacred sites; visitors are now requested not to make the climb, but are not prevented from doing so.

Wet Tropics of Queensland

The process leading up to the nomination of this property involved at times heated conflict between the Federal and Queensland governments, and with other stakeholders, particular those in the timber industry. This conflict culminated in a 1988 High Court challenge by Queensland against nomination. The case was dismissed, as was another in 1989 challenging the proclamation of the property under the *World Heritage Properties Conservation Act (Cwlth) 1983*. Situated between Townsville and Cooktown, this is another property inscribed under all four natural criteria. It includes spectacular scenery and rugged topography, most of the remaining tropical rainforest in Australia, many rare or endangered species of plants and animals, and an almost complete record of the evolution of plant life on earth. The property continues to hold great significance for Indigenous peoples. Covering almost 900,000 ha, it is not a continuous area, as there are many gaps where cleared and unreserved land separates its component parts. The Wet Tropics Management Authority, established under complementary Commonwealth and State legislation, provides overall strategic management, whereas day-to-day management is carried out by a number of Queensland agencies, notably QPWS. There were extensive WHC discussions relating to this property from its inscription to 2001, generally about the implementation of management structures, a management plan, and monitoring processes. The WHC response was ultimately very positive. The Australian Government, in its 2003 report, perceived internal fragmentation to be a major issue, even though significant areas formerly in private ownership had been incorporated into protected reserves. Obsolete infrastructure had been phased out and management and maintenance practices improved.

Shark Bay, Western Australia

Shark Bay's inscription was justified on the grounds of it representing major stages in the earth's evolutionary history, being an outstanding example of on-going ecological and biological processes, being an example of superlative natural phenomena, and containing habitats of biodiversity conservation significance. It is thus yet another of the rare properties inscribed under all four natural criteria. In addition to the sea-grasses, dugongs and stromatolites, the area is important for a rich avifauna, 26 species of endangered mammals, large numbers of species of amphibians and reptiles, humpback whales, green and loggerhead turtles, and dolphins. The WA Government, formerly through the Department of Conservation and Land Management (CALM) and now Department of Environment and Conservation, is responsible for day-to-day management, but overall policy is co-ordinated by a Ministerial Council of two State and two Federal ministers, acting under a State–Commonwealth Agreement and joint legislation. While the majority of the area is in various conservation reserves, some is in private ownership. When inscribed, the WHC was concerned by the lack of a State–Commonwealth agreement and adequate management structures, but these shortcomings were rectified over the next few years. In 1998, concern was expressed by the WHC at the granting, by the WA Government, of a petroleum exploration licence for an area within the property. The Australian Committee of the IUCN prepared and presented a comprehensive report on the property's conservation to the WHC in 1999; the Australian and WA governments were, in 2000, proceeding to use this report and its recommendations as a basis for a strategic plan for the property. By that stage, any petroleum exploration or production activity came under the EPBCA and would need Commonwealth approval under strict guidelines. As with most World Heritage properties, increasing visitor numbers have the potential to have detrimental impacts.

Fraser Island

This property was inscribed because it represents significant on-going ecological and biological processes and is an example of superlative natural phenomena. The massive sand deposits give a continuous record of climatic and sea level changes over 700,000 years, while the island has areas of exceptional beauty and important flora and fauna. It has a long history of Aboriginal occupation, although it was not inscribed on cultural grounds. Most of the island is within the Great Sandy National Park and managed by QPWS. Visitor access and development proposals are tightly managed, although conservationists have some concerns on both counts. The WHC expressed some concern in 2001 because the island did not have its own plan of management, but, rather, came under the Great Sandy Region Management Plan that also included Cooloola National Park on the mainland. That arrangement was to be reviewed by Australia.

Australian Fossil Mammal Sites (Riversleigh/Naracoorte)

This is an unusual site as its two components are over 2,000 km apart: Riversleigh is in north-western Queensland (in Lawn Hill National Park) and Naracoorte is in south-eastern South Australia (in Naracoorte Caves National Park). Each site is owned and managed by the relevant state government. Fossils found at Riversleigh have been crucial in developing understanding of Australia's mid-Cainozoic vertebrate diversity and include material from 15 million years ago. Victoria Cave at Naracoorte contains the country's largest, best-preserved, and most diverse deposits of Pleistocene vertebrate fossils. Both sites are important for on-going scientific research. According to Australia's 2003 report, the main challenges at Riversleigh relate to enhancing protection at the remote site and establishing a community consultative process that includes Indigenous people, among others. IUCN had reported the problems of vandalism and theft at Riversleigh, exacerbated by the lack of infrastructure and a ranger presence, to the WHC in 2001. At Naracoorte, mitigating impacts of visitors and improving interpretation are the key issues.

Heard and McDonald Islands

Justification for inscription was because the property contains outstanding examples of major stages in the earth's history and development, and significant on-going ecological and biological processes. Big Ben is a 2,745 metre-high active volcano covered in snow and glacial ice. While the diversity of life is low, there are huge numbers of the species present, including seals, penguins and flying sea birds, and no species known to have been introduced by humans. The initial nomination in 1991 was deferred because of uncertainty at the WHC as to comparisons with other subantarctic islands, and because of a desire for further information. When the WHC inscribed the property in 1997, it requested a report from Australia on the marine resources of the surrounding seas. After extensive research by the Australian Antarctic Division, this report and a plan to establish a Marine Protected Area were submitted in 2000. The Heard and McDonald Islands Marine Reserve was declared under the EPBCA in October 2002.

Macquarie Island

This property was inscribed because it provides a unique example of exposure of the ocean crust above sea level, and evidence of sea-floor spreading. It was accepted as being an outstanding example representing earth's evolutionary history in terms of on-going geological processes, and as containing superlative natural phenomena. It is a Strict Nature Reserve under the IUCN categorisation and has been under various forms of reservation since 1933, but while the island also has important fauna and flora, it was not inscribed for biodiversity reasons. The land area is in a State Reserve (of Tasmania), and the marine values are protected within the Macquarie Island Marine Park, declared in 1999, and the world's largest highly protected marine area. Access to the island is strictly limited, almost entirely to scientific expeditions. Initially nominated in 1992 as the Macquarie Island Nature Reserve, the property was eventually inscribed for its geological, rather than biological, significance in 1997 after a referral back to Australia and resubmission. The WMC encouraged Australia to pursue a joint nomination with New Zealand to incorporate a number of their subantarctic islands, and the question of a future renomination on biological criteria was left open.

The Greater Blue Mountains Area

Inscription was justified on the basis of the diversity of eucalypt habitats, including wet and dry sclerophyll, mallee heathlands, and localised swamps, wetlands and grassland. Over 13 per cent of all eucalypt taxa are represented, including all four major groups, and there is a high level of endemism, with 114 endemic taxa and 120 nationally rare and threatened plant taxa. When the nomination was presented in 1999, the IUCN recommendation was against inscription, but encouraged development of a serial nomination including other eucalypt-related sites. Many people live in the Blue Mountains, even if not technically within the property, and more than three million people visit each year, so human impacts are great. The property is protected within national parks and other reserves, and overall management is by the NSW NPWS. Cultural criteria are referred to in reports of the 1999 WHB and WHC meetings, but the Bureau recommended against inscription on such grounds; it appears that cultural reasons were omitted from the revised nomination considered in 2000. The property was inscribed in 2000 as a stand-alone, the 1994 discovery of the unique Wollemi pine quite possibly influencing the positive outcome. Australia assured the WHC that other eucalypt sites could be registered on the newly introduced National Heritage List.

Purnululu National Park

The property was inscribed on the basis of outstanding universal geological value as the most outstanding example of cone karst in sandstones in the world, and because of the recent international recognition of the area's exceptional natural beauty, especially in the banded, beehive-shaped cone towers of the Bungle Bungle Range. Landforms and ecosystems elsewhere in the Park provide a visual buffer for the central features, even though not of major significance themselves. The property had been nominated by Australia as a site also satisfying cultural criteria C(iii), C(v) and C(vi) on the basis of on-going traditional Aboriginal occupation, but the property was not accepted as either a mixed site or a cultural landscape. Even though the ICOMOS recommendation was that the decision be deferred to allow Australia to provide further information, their report to the WHC also stated that the site should only be inscribed as a mixed site because of the inseparability of its natural and cultural attributes. Concern was expressed by the WHC, IUCN and ICOMOS that the cultural attributes had been diminished over time, but the Australian Government intends to pursue the cultural element of its nomination, perhaps seeking a conversion to a cultural landscape listing. Purnululu has a relatively short conservation history, being established as a national park in 1987 and managed by CALM/DEC, with Traditional Owners being involved more and more in its management. Previous overgrazing in the lower areas surrounding the Bungle Bungles is a major problem, and if visitor numbers grow significantly, careful control of visitor pressures will be essential. At the 2005 WHC meeting, concerns were expressed about conservation issues, the need to sustain local Indigenous communities, and the staffing and financing of this remote site.

Royal Exhibition Building and Carlton Gardens

Inscription was justified by the fact that this property constituted the main extant survivor of a Palace of Industry and its setting, reflecting the global influence of the international exhibition movement of the nineteenth and early twentieth centuries. Few such properties remain, and even fewer retain their authenticity in terms of original location and condition, while the Great Hall, built for the 1880 Melbourne International Exhibition, is the only substantially intact example remaining from a major international exhibition. This was the first, and so far only, non-indigenous cultural site in Australia to gain World Heritage status. The Royal Exhibition Building is owned by the Victorian Government and administered and managed by the Museums Board of Victoria, while the Carlton Gardens are managed by the City of Melbourne. The property is protected under Victorian legislation as well as, since inscription, under the Commonwealth's EPBCA.

Actual and possible future nominations

All State Parties are expected to have Tentative Lists, and following recent changes in the Operational Guidelines properties can only be nominated if they are on such a list lodged with the World Heritage Centre. Australia had only two entries on its Tentative List in early 2006, these being the Sydney Opera House and Australian Convict Sites. The former was nominated in 2005 and will be considered for inscription at the June-July 2007 WHC meeting. The Lake Eyre Basin and Cape York Peninsula are possible future additions, as are additional eucalypt-related sites.

Sydney Opera House

'Sydney Opera House in its Setting with the Sydney Harbour Bridge and the Surrounding Waterways of Sydney Harbour from Bradleys Head to McMahons Point' was originally nominated in 1981. The Bureau decided that 'modern structures should only be accepted when there was clear evidence that they established, or were outstanding examples of, a distinctive architectural style', and suggested a revised nomination that focused on the Harbour and had the Opera House and Bridge as incidental elements, not primary features. Australia withdrew its nomination before the WHC meeting later in 1981. The site was entered on the Tentative List in 1996 and an unpublished nomination under the title 'Sydney Opera House in its Harbour Setting' produced, but drawn-out discussions between the Federal and NSW governments delayed a new nomination. The 2005 nomination concentrates even more on the building itself than did either the 1981 or 1996 versions, although the new one does include as a buffer zone the portion of the Harbour mentioned earlier. The Australian Government seems to be banking on two facts: the WHC has in recent years been more open to inscribing modern architecture; and the importance of the Sydney Opera House has become much more widely accepted over the intervening years. The present nomination is under criterion C(i) alone. Some brief extracts from the nomination document are as follows:

> The Sydney Opera House is a work of human creative genius, and a masterful architectural and engineering achievement ... It was a turning point in the late Modern Movement, a daring and visionary experiment resulting in an unparalleled building that defied categorisation and found an original style in which to express civic values in monumental public buildings. The influences that resulted in the Sydney Opera House's unique form include organic natural forms and an eclectic range of aesthetic cultural influences, brilliantly unified in the one sculptural building ... Moreover, the Sydney Opera House's significance is intrinsically tied to its harbour-side site (Australian and NSW Governments 2006, 27).

Australian Convict Sites

No information on this potential nomination is available to the public, but it is believed to contain at least portions of The Rocks in Sydney and Battery Point in Hobart, and possibly parts of Fremantle. Many other smaller sites are possible additional components.

Conclusion

Australia has been heavily involved in the World Heritage regime ever since the World Heritage Convention first came into force, and has played a prominent role in the various processes and associated bodies involved. There are now sixteen World

Heritage sites in Australia (Table 3.3), all of which are basically well managed to maintain their World Heritage values. On the other hand, there have been, and continue to be, development pressures that have the potential to impinge on those values; many of these have been discussed by the WHB and WHC, and some are mentioned in this chapter. The most controversial of these issues involved actual and potential uranium mining in enclaves 'within' Kakadu National Park, and this matter led to a definite souring of the relationship between the Australian Government and the WHC and associated bodies. One of the most contentious aspects was the relativity between Australia's national sovereignty and its international obligations under the Convention, including the right of the WHC to intervene and, especially, its right to add Kakadu to the LWHD against Australia's wishes. That tension has not been resolved, but merely placed on the back burner. Despite that, Australia continues its active involvement, and two additional sites have been inscribed on the WHL since the height of the Kakadu controversy, while another is presently awaiting a WHC decision. The last successful nomination and the current one indicate a shift in emphasis from natural and Indigenous sites to non-indigenous cultural sites. Australia, however, continues to have a list of World Heritage sites much more heavily weighted toward natural sites than that of almost any other State Party.

References

Aplin, G. (2004) 'Kakadu National Park World Heritage Site: deconstructing the debate, 1997–2003', *Australian Geographical Studies* 42:2, 152–74.

Australia (2003) *World Heritage Periodic Reporting Asia–Pacific Region, 2003: Australian Contribution to the Regional Synthesis Report, December 2002* (Canberra: World Heritage Branch, Department of the Environment and Heritage).

Australian and NSW Governments (2006) *Sydney Opera House: Nomination by the Government of Australia for Inscription on the World Heritage List 2006* (Canberra: Department of the Environment and Heritage).

Environment Australia (1999) *Australia's Kakadu: Response by the Government of Australia to the UNESCO World Heritage Committee Regarding Kakadu National Park* (Canberra: Environment Australia).

UNESCO (1998) 'Executive summary and list of recommendations', in *UNESCO Report on the Mission to Kakadu National Park, Australia*, Information Document for the World Heritage Bureau Meeting, 29 November 1998, Kyoto (Paris: World Heritage Centre, UNESCO).

Websites

UNESCO: http://www.unesco.org/
World Heritage Commission/Committee/Convention: http://whc.unesco.org/
List of World Heritage properties: http://whc.unesco.org/en/list/
MAB Biosphere Reserves: use UNESCO site and search for MAB

IUCN: http://www.iucn.org/

ICOMOS: http://www.icomos.org/

Ramsar Convention: http://www.ramsar.org/

Biodiversity Convention: http://www.iisd.ca/process/biodiv_wildlife.htm

UNEP World Conservation Monitoring Centre: search for 'wcmc' in http://www.unep.org/

World Commission on Protected Areas (of IUCN): http://www.iucn.org/themes/wcpa

Environment Australia heritage site: http://www.deh.gov.au/heritage/index.html

Biodiversity pages, DEH: http://www.deh.gov.au/biodiversity/index.html

Australia's World Heritage: http://www.deh.gov.au/heritage/worldheritage/ (there are further links for each site)

New legislation, EPBC Homepage: http://www.deh.gov.au/epbc/index.html

Australia ICOMOS: http://www.icomos.org/australia/

Asia-Pacific Focal Point for World Heritage Managers: http://www.heritage.gov.au/apfp/

Chapter 4

The Changing Geographies of Australia's Wilderness Heritage

C. Michael Hall

Images of the 'bush' and the 'outback' are now an essential element of Australia's wilderness experience and a fundamental contributor to Australia's cultural identity. Wilderness also serves as a source of cultural myth and representation, both in terms of natural images, i.e. rainforest, desert and reef and the animals that inhabit them, as well as cultural identities and stereotypes, i.e. the 'ocker', the 'man from Snowy River', Ned Kelly, the 'wild colonial boy', and the 'digger'. Contemporary media, and particularly film and advertising, has also served to reinforce the myth of the bush in Australia's national identity, even though it is one of the most urbanised countries in the world (Waitt 1997).

Commodified representations of Australia's wilderness experience have also become integral elements in the promotion of Australia to attract tourists, migrants and even investment (Hall 2007). In the 1980s 'Crocodile Dundee' related representations were utilised extensively by the Australian Tourist Commission and arguably still are an important element within contemporary national tourism advertising. For example, the Australian Tourist Commission (ATC) Brand Australia campaign launched in 2001 utilised 'traditional' images (ATC 2001a, b) that according to the ATC (2001a): 'positions Australia as a friendly, colourful and stylish destination and will be seen by 300 million people in 11 countries'. ATC research (2001a) indicated: 'that the free-spirited, Aussie personality is one of the most powerful assets we have ... The design was developed after consumer testing identified the kangaroo as the country's most recognisable symbol, while the colour variations represent the diversity of the coastal and interior climates of Australia'.

Many of these 'traditional' images remain used in the 2006 'Australian invitation' campaign (otherwise better known as 'where the bloody hell are you?') (Hall 2007). According to Tourism Australia (2006) the brand insight of the campaign is that 'Australia has a uniquely open personality which characterises its experiences' with the Australian character/personality being the point of differentiation. For Tourism Australia (2006, 6), 'The campaign presents a single and compelling brand message that speaks to the common motivations of our target consumer in every market' although 'specific executions and combinations of executions have been tailored for each market'. Indeed, for some parts of Australia, such as the Kimberley in Western Australia, in keeping with the Brand Australia strategy, are promoted as the 'last frontier' in which the tourist can come into contact with wilderness (e.g. Waitt and Head 2002; Waitt et al. 2003). The contemporary commodification of nature and outback identity for the purposes of conveying a positive message to attract visitors is perhaps ironic given the history of wilderness in Australia and changing perceptions and values.

This chapter will provide an historical account of the changing geographies of wilderness as heritage. Before charting the changing geographies of Australia's wilderness heritage the chapter will briefly discuss the wilderness concept. Several themes are then identified including the growth of an Australian wilderness conservation ethic and movement; the growth in Commonwealth regulatory capacity in addition to state responsibilities and, more recently, to international heritage regimes, such as the World Heritage Convention; and the ongoing importance of the economic rationale of tourism as a justification for wilderness conservation.

The Wilderness concept

Wilderness is a concept with many layers of meaning. Indeed, Tuan (1974, 112) went so far as to argue that 'wilderness cannot be defined objectively: it is as much a state of mind as a description of nature'. The problem of defining wilderness was well summarised by Nash (1967, 1) who observed that '"Wilderness" has a deceptive concreteness at first glance. The difficulty is that while the word is a noun it acts like an adjective. There is no specific material object that is wilderness. The term designates a quality (as the '-ness' suggests).' Indeed, the idea of wilderness is primarily determined from the northern European experience of nature (Oelschlaeger 1991) in which places 'of wild beasts' were landscapes of fear outside of the boundaries of 'civilisation'. Such a perspective has been highly influential in determining not only how wilderness areas have been perceived but also how they might be conserved, with for example the 1964 US Wilderness Act defining wilderness as 'an area where the earth and its community of life are untrammelled by man, where man himself is the visitor that does not remain'. The four defining qualities of wilderness areas protected under the Act are that such areas:

1. generally appear to be affected by the forces of nature, with the imprint of man substantially unnoticeable;
2. have outstanding opportunities for solitude or a primitive and unconfined type of recreation;
3. have at least 5,000 acres (2,023 hectares), or be of sufficient size as to make practical its preservation and use in an unimpaired condition;
4. may also contain ecological, geological or other features of scientific, educational, scenic or historical value.

From a conservation perspective this has not only meant the designation of large areas of land under legislative protection but, historically, also reinforced the perspective that land that had been used by 'unsettled' indigenous peoples could qualify as wilderness. The drive for wilderness conservation in the United States and, more recently in Australia and elsewhere around the world, had its origins in the desire for wilderness recreation experiences as opposed to the inherent biological conservation values of such areas that were only recognised much later (Hall 1992). This has meant that there are fundamentally two conceptions of the qualities of wilderness. One is anthropocentric, in which human needs including recreation are considered to be paramount. The other biocentric (or ecocentric) approach defines wilderness essentially in ecological terms and equates wilderness quality with a relative lack of human impact (Hall and Page 2006).

In Australia there has been a shift from anthropocentric to biocentric approaches in defining wilderness quality (see Hall and Page 2006). Kirkpatrick's and Haney's (1980, 331) study of south-west Tasmania identified wilderness as 'land remote from access by mechanised vehicles, and from within which there is little or no consciousness of the environmental disturbance of western man'. Kirkpatrick and Haney assigned absolute wilderness quality scores based on the more readily quantifiable characteristics of wilderness: remoteness and primitiveness. These characteristics are the two essential attributes of wilderness which fulfil biocentric and, potentially, anthropocentric perspectives on wilderness. The relative attributes of remoteness and primitiveness can be expressed as part of a continuum that indicates the wilderness quality of a region (Helburn 1977; Hall 1992) (Figure 4.1). Remoteness is measured 'as the walking time from the nearest access point for mechanised vehicles', whereas primitiveness, which 'has visual, aural and mental components', is 'determined from measures of the arc of visibility of any disturbance ... and the distance to the nearest disturbance' (Kirkpatrick and Haney 1980, 331). The identification of remoteness and primitiveness as the essential attributes of a wilderness area helped provide the basis for the national survey of wilderness that was supported by the Commonwealth government during the 1980s and 1990s (Lesslie and Taylor 1985; Lesslie 1991; Lesslie and Maslen 1995).

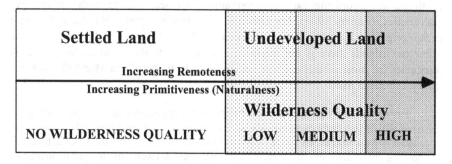

Figure 4.1   The wilderness continuum

Source: Hall 1992

Importantly, the Australian experience with wilderness inventory indicated that there has been an 'almost universal tendency to confuse the *benefits derived from wilderness* with the *nature of wilderness itself*' (Lesslie and Taylor 1983, 14) – a point of crucial importance in the delineation, inventory, and management of wilderness. Hence, the two attributes which are definitive of wilderness, *remoteness* (distance from the presence and influences of settled people) and *primitiveness* (the absence of environmental disturbance by settled people) need to be based at the high-quality end of the wilderness continuum to accommodate the anthropocentric and biocentric dimensions of wilderness (Hall and Page 2006). In the Australian National Wilderness Inventory four indicators of wilderness quality were identified: remoteness from settlement, remoteness from access, aesthetic primitiveness (or naturalness) and biophysical primitiveness (or naturalness), and were used to generate a series of maps of wilderness quality in Australia. However, the identification of wilderness does not necessarily translate into the conservation of such areas (Herath 2002; Hall and Page 2006). Instead, this requires political action that is, in turn, grounded in values and interests, and it is these that have changed substantially over the past 200 years.

Conserving Australia's wilderness

The evolution of wilderness preservation in Australia has distinct parallels with other colonial new world countries in North America and New Zealand and has been particularly influenced by the United States experience. The European settlers' encounter with the Australian environment contained Romantic elements akin to those which operated in Canada and the United States, while the rise of the progressive conservation movement in the United States also had a major influence on Australian attitudes towards conservation of the natural landscape (Powell 1976; Hall 1992). The themes that national parks were worthless lands and the importance of tourism, so central to the development of national parks in North America and New Zealand, were also repeated in the Australian context.

The European settlers of Australia were faced with a new world, which was to them, as a contemporary commentator described, replete with 'antipodean perversities' (Finney 1984). For example, the French explorer Baudin was aghast at the 'primitive' nature of the western coast of New Holland. 'In the midst of these numerous islands there is not anything else to delight the mind ... the aspect is altogether the most whimsical and savage ... truly frightful' (in Marshall 1968, 9). To the European mind

> ... rare conservatory plants were commonplace; the appearance of light-green meadows lured settlers into swamps where their sheep contracted rot, trees retained their leaves and shed their bark instead, the more frequent the trees, the more sterile the soil, the birds did not sing, the swans were black, the eagles white, the bees were stingless, some mammals had pockets, others laid eggs, it was warmest on the hills and coolest in the valleys, even the blackberries [wild raspberries] were red, and to crown it all the greatest rogue may be converted into the most useful citizen: such is *Terra Australia* (J. Martin 1838, in Powell 1976, 13–14).

Despite the adverse initial reaction to the Australian landscape, attitudes were not always unfavourable. Instead, a generally ambivalent perception of the Australian environment gradually emerged although the Europeanisation of the Australian environment was for a long time a goal of many settlers. However, one of the main differences between the beginning of European settlement in Australia and in North America, and therefore between attitudes towards nature, was determined by the periods during which initial settlement took place. The 'howling wilderness' (Nash 1967) of the New England coast and the eastern seaboard of North America was settled before the emergence of a favourable aesthetic reaction to wild places in European intellectual thought. In contrast, the first waves of European settlement in Australia occurred during a period in which more favourable attitudes towards wild nature were developing. Nevertheless, it is important to recognise that the first national parks and conservation reserves in Australia, as in the United States, were conserved because the land was worthless for traditional forms of economic development such as agriculture, mining and forestry with economic value seemingly only being able to be extracted via the scenic and recreational values of recreation and tourism. As Runte (1983, 138) observed, 'everyone would prefer to attribute the national park idea to idealism and altruism'. Runte's 'worthless lands' argument arose from the very first speech in Congress that contained elements of the national park idea. Senator John Conness of California, on introducing a bill to cede Yosemite to the State of California as a park noted, somewhat paradoxically, that the location in the Sierra Nevada mountains was 'for all public purposes *worthless*, but which constitute, perhaps, some of the greatest wonders in the world' (Congressional Globe, 38th Congress, 1st session, May 17, 1864, 300 in Runte 1979, 48-49). However, the creation and continued protection of national parks in Australia, as well as in the United States and elsewhere, has been as much dependent on the absence of material wealth as it has been on the weight of aesthetic and ecological arguments.

Hall (1992) argued that the worthless lands hypothesis applies to Australia as well as to the United States. A glance at a map of Australia's and New Zealand's national parks and reserves system reveals that reserves are primarily located at past or present frontiers of economic development. For instance, the vast majority of Western Australia's reserves are located in the arid inland (Pouliquen-Young 1997) along with the majority of Aboriginal reserves.

The first Australian 'national park' was established in New South Wales. In 1879, seven years after the creation of Yellowstone National Park in the United States (and 22 years before Australian Federation), 18,000 acres (7,284 hectares) of land were set aside as a National Park at Port Hacking, 22 km south of Sydney's city centre. This area was increased to 14,000 hectares the following year. An exhibit organised by the Royal Society of New South Wales in 1878 contained a description of Yellowstone but it is unlikely that the American National Park provided much more than a name for the new park (Slade 1985–86). Instead, the creation of the National Park (later Royal National Park) was inspired more by a desire to ensure the health of Sydney's working population than to provide a remote scenic wilderness experience as per Yellowstone, although it should be noted that the health benefits of outdoor recreation, particularly for males who may then serve in the armed forces, was also a factor in United States park development (Nash 1967). According to a member of the New South Wales Legislative Assembly, John Lucas, the park was created 'to ensure a healthy and consequently vigorous and intelligent community ... all cities, towns and villages should possess places of public recreation'; while Sir Henry Parkes commented, 'The Honourable Member says it is a wilderness and that years must elapse before it can be of any use, but is it to remain a wilderness? ... certainly it ought not to remain a wilderness with no effort whatever to improve it'

(1881, in Mosley 1978, 27). Indeed, according to Pettigrew and Lyons (1979, 17), the area reserved was available only as 'a consequence of the poor quality of much of it and of the Georges River between it and the expanding Sydney'.

The reasons for the establishment of Australia's first national park are similar to those pertaining in Canada and the United States. First, there was no cost to the government in the reservation of land as it was already held under state control. Second, the land was regarded as worthless with no value for agriculture, although timber cutting and grazing were allowed to continue in the park until well into the twentieth century. Third, the development of a railroad line enabled Sydney's inhabitants to travel to the Park. However, in contrast to the American situation, the park was established to provide for mass recreation rather than the elite commercial recreation that characterised the early days of Yellowstone, although commercial hotels were built in the park soon after it was established. In addition, the landscape value of the National Park was related to the coast and rivers rather than mountain scenery or spas as in the United States. Nevertheless, as in America, the area was 'improved' with suitable types of development such as military parade grounds, picnic areas, bandstands, and zoological displays. Despite these initial 'improvements', many of which have been removed in the age of ecology, the park has become a significant component of the national park system of New South Wales.

The New South Wales experience was repeated throughout Australia. For example, the first national park established under the Queensland *State Forests and National Parks Act* of 1906 at Tambourine Mountain was on land that was judged, according to Powell (1976, 114), as 'unfit for any other purpose', although the tourism benefits were noted. In the case of the passing of legislation 'to provide for the reservation, management and protection of ...national parks' in Queensland in 1906 the Hon. J.T. Bell, Secretary for Public Lands commented on the significance of 'areas which ... as localities are likely to become popular resorts as the population grows larger – places to which those who desire to take a holiday may like to go from time to time and know that they will get pure air, good scenery and country life' (1906, in Goldstein 1979, 133–4).

Unlike the United States, which had a well-developed national wilderness conservation movement by the turn of the twentieth century, Australia did not have any wilderness specific interest groups until 1932 when the National Parks and Primitive Areas Council (NPPAC) was established in New South Wales. Like the Sierra Club in the United States the NPPAC grew out of walking clubs. However, unlike the Sierra Club, the NPPAC was primarily state focussed whereas the Sierra Club had a multi-state, if not national agenda with respect to national park creation and wilderness conservation.

Amongst the objectives of the NPPAC was the advocacy of 'the protection of existing tracks, paths and trails in use, particularly those having scenic and historical interests and values' (Bardwell 1979, 17). The NPPAC were strongly influenced by American conservation initiatives (Strom 1969). For example, in 1932 Myles Dunphy, the leader of the NPPAC, obtained a supply of booklets on American national parks which served as propaganda for the national park idea in Australia (Thompson 1985), and doubtless influenced the way in which the bushwalking movement and the NPPAC approached campaigns for the preservation of natural areas in Australia.

Indeed, the American influence in New South Wales national parks conservation was historically so strong that when the first national parks agency was established in the 1960s it replicated the American national park service even down to the naming of certain positions (Hall 1992).

Up until the emergence of the NPPAC, proposals for national parks and wilderness conservation had essentially been local in nature with support from key individuals in the political and scientific elites. The NPPAC's Greater Blue Mountains National Park Scheme probably represented the first major attempt of an Australian conservation group to mobilise mass support for the preservation of wilderness. On 24 August 1934, the NPPAC paid for a four-page supplement, complete with maps and photographs, to be included in the local *Katoomba Daily*. The supplement was highlighted by Myles Dunphy's proposal for a Blue Mountains National Park with 'primitive areas', an American term of the 1920s and 1930s that was used with respect to the identification of wilderness within the US Forest Service:

> The Blue Mountains of Australia are justly famous for their grand scenery of stupendous canyons and gorges, mountain parks and plateaux up to 4,400 feet altitude, uncounted thousands of ferny, forested dells and gauzy waterfalls, diversified forest and river beauty, much aloof wilderness and towns and tourist resorts replete with every convenience for the comfort and entertainment of both Australian and overseas visitors (National Parks and Primitive Areas Council (NPPAC) 1934, 1).

That the supplement attempted to link the scenic attractions of the area with tourism is hardly surprising. Australia was then in the grip of a depression, and linking preservation with positive economic benefits was logical. However, it is also interesting to note that the NPPAC (1934, 1) argued that the sandstone country of the Blue Mountains 'is potentially desert land', thereby reinforcing the 'worthless' lands concept of wilderness. Although the bushwalking groups and the NPPAC did much to raise awareness of national parks in the general population their overall political effect in conservation terms was rather limited and localised. Instead, it would take until the 1960s and a greening of Australian politics for a more effective conservation movement to emerge in Australia.

The 1960s saw the birth of environmental awareness and environmental lobby groups on the world stage. The publication of books such as *Silent Spring* and the images associated with the *Torrey Canyon* oil disaster in Britain did much to raise awareness of the need for environmental protection. In addition, the European vision of the Australian landscape was gradually coming to be replaced by a more sympathetic Australian sense of place, which started to value the Australian environment (Seddon and Davis 1976). While the various Australian states had declared national parks on a piecemeal basis, it was not until the late 1960s and early 1970s that state national park systems with parks under a single park management agency came into existence, along with the first state environmental protection authorities. Perhaps, more significantly, it was also at this time that parks began to be declared in areas which had high ecological and wilderness values even though they also had other economic values in terms of minerals or timber. For example, the Great Barrier Reef came to be declared a marine park due to conservation group concerns over oil drilling. Similarly, conservation groups lobbied to stop mineral sands development on Fraser Island, which, like the Great Barrier Reef, is now a World Heritage Area (see Aplin, Chapter 3). In both cases arguments for the conservation of nature for its intrinsic value were entwined with economic conservationist arguments that national parks should be established because of their value for tourism (Wright 1977; Hall 1992).

Further complicating national park issues in Australia, was the gradual strengthening of Commonwealth Government powers with respect to the environment and national parks under the reformist Whitlam Labor Government (1972–1975). The development of Commonwealth legislation provided a mechanism by which conservation groups could seek to override state government inaction or recalcitrance in conserving natural areas through national park declarations. Indeed, Australia's signing of the World Heritage Convention in this period provided the capacity for Australia's conservation debates to become international in scope in the 1980s and 1990s. However, it should be noted that the implementation of the World Heritage Convention in Australia has more often been a debate over issues of state rights rather than the creation of an effective management regime to preserve World Heritage values (Hall 1992, 2006; and see Aplin Chapter 3). In fact one of the ironies surrounding national parks and wilderness in Australia, is that under the Australian constitution it is the states that have primary responsibility for land use, therefore few national parks are actually under effective national control.

Other countries, such as Canada and the United States, have overcome this situation through land purchase federal-state agreements. This has not occurred in Australia despite moves in this direction. For example, in 1981 the House of Representatives Standing Committee on Environment and Conservation recommended that,

> some areas of national and international significance should be administered by the Commonwealth as truly 'National' national parks under Commonwealth legislation … Areas so declared would be more likely to receive more appropriate resources, and to be administered and protected in the national interest, free from purely local or state pressures. We would envisage only a small number of these parks, but that as a group they would represent outstanding areas of Australia's natural heritage (1981, 37).

Perhaps ironically the primary reason why such a system of truly national parks has not developed was because of the role of wilderness conservation and environmental issues as key items on the political agenda in the 1980s and 1990s. In this period, conservation issues, such as those relating to the Franklin Dam and the Rainforests of Queensland, became national environmental issues that emerged as key concerns in federal elections and politics. National interest groups such as the Australian Conservation Foundation and The Wilderness Society (formerly the Tasmanian Wilderness Society) exerted considerable influence on federal politics. Using its external affairs and corporations powers under the Australian Constitution, the federal government was able to use its World Heritage Properties Conservation Act to enforce Australia's obligations under the World Heritage Convention and control land use in areas of World Heritage value (Hall 2006). Such a situation created enormous tensions over state rights, as well as concerns over the extent to which areas with high natural values, such as wilderness, could be used by the Commonwealth to prevent various developments on such lands. In this type of political environment it was therefore politically impossible to develop a national park system as envisaged by the Standing Committee.

The election of a Coalition federal government in 1996 represented a substantial shift in Australian wilderness politics. In November 1997 a Heads of Agreement on Commonwealth and State Roles and Responsibilities for the Environment (Council of Australian Governments 1997) was signed by all heads of federal and state government and by the Australian Local Government Association. A new Act, the *Environment Protection and Biodiversity Conservation Act 1999*, governing the Commonwealth's responsibilities with respect to World Heritage, as well as other significant environmental matters of national interest came into effect from 16 July 2000. Perhaps, most significantly, wilderness has not been a significant national political issue since the mid-1990s and the Howard government has demonstrated no interest in seeking to conserve Australian wilderness areas that had been identified in the national wilderness inventory. Instead, such responsibilities have been seen as a state responsibility, with legal wilderness protection being considered under state national parks legislation and management plans. Furthermore, it may also be the case that the conservation 'victories' with respect to the stopping of the Franklin Dam, retention of old-growth temperate forest, and rainforest conservation during the Labor years in power may mean that many people believe that Australia's wilderness is now protected (Mulligan 2001)

However, this is not the case. As the frontier of economic development and environmental exploitation advances, even the more remote wilderness areas and parks and reserves are threatened by material interests In the rush to 'open up' economically peripheral areas, ecology and aesthetics are secondary considerations in the decision making process, a situation that has been long recognised (Hall and Boyd 2005). As the Committee of Inquiry into the National Estate (1975, 77-78) reported: 'National parks and other large reserves have generally been made only in areas *unwanted for any other purpose. Sectional pressures have ensured that other areas, whether their potential is for agriculture, grazing, mining, forestry, water storage or settlement, have largely remained unreserved'* [author's emphasis]. One of the few rigorous studies of the representativeness of biodiversity conservation in protected area systems in Australia has been undertaken in Tasmania (Mendel 2002; Mendel and Kirkpatrick 2002). In their study Mendel and Kirkpatrick (2002) reported that before the 1970s the representation of plant communities in the Tasmanian national park system was strongly biased toward the reservation of communities that were not economically viable. In 1970, less than 3 per cent of economically valuable communities were reserved above 15 per cent of their pre-European value, compared with over 17 per cent for those without economic value. However, as a result of additions to the reserve system by the Tasmanian government, by 1992 one-third of the plant communities in Tasmania had 15 per cent of their pre-European area included within the state national parks and reservation system. This included over 20 per cent of economically valuable communities and 58 per cent of non-economically valuable plant communities (Mendel and Kirkpatrick 2002). According to Mendel and Kirkpatrick (2002) their research supports the 'worthless lands' hypothesis. However, they also note that despite some communities being unrepresented, for example, treeless lowland communities, the Tasmanian reserve system is closer to a representative system of biodiversity than those of most countries.

The Tasmanian experience with wilderness, for much of the last 40 years the focal point for wilderness conservation in Australia, provides a good example of the increased complexities facing Australia's wilderness heritage. First, tourism remains an integral economic use of wilderness that provides an economic justification for its conservation (Kirkpatrick 2001). Second, there is increased recognition that wilderness constitutes a significant cultural heritage as well as natural heritage through its capacity to maintain biodiversity. Initially, this was recognised through Aboriginal relationships with the land (e.g. Adams 2004), but increasingly it is also being recognised that Australia's European settlers have significant cultural relationships to wilderness areas that need to be acknowledged in wilderness management strategies (Russell and Jambrechina 2002). Finally, Australia's wilderness heritage is essentially residual land, that is land unwanted for other purposes, with consequent implications for the management of biodiversity (Sattler and Creighton 2002), particularly under conditions of environmental change as well as Aboriginal land claims.

Conclusions

For many Australians the country's wilderness heritage probably seems secure. In the issue-ecology of policy in Australia, other environmental issues, and particularly climate change, now dominate the government and, to an extent, the public agenda (Hall and Higham 2005; Gössling and Hall 2006). Yet wilderness is actually vital to climate change, not only through its function as a carbon sink in many cases, but also because of its role as evolutionary and ecological refugia. The loss of biodiversity as a consequence of anthropogenic induced environmental change therefore actually makes the biological conservation of Australia's wilderness heritage even more important even as wilderness quality diminishes. While, for many peripheral areas, nature-based tourism offers one of the few alternatives for economic development (Hall and Boyd 2005; Hall and Härkönen 2006). New interpretations of wilderness conservation also mean that, arguably for the first time in Australia's landscape conservation history, maintenance of the wilderness values of private land is also being given far greater consideration (Knight 1999; Syder and Beder 2006) with corresponding implications for understandings of what constitutes heritage.

The perception of the value of Australia's wilderness has changed over time and continues to change. Australia's wilderness has generally come to be seen in an increasingly positive light particularly given its role in the development of national myths as well as representations of Australia both to Australians and to the world. Yet the cultural role has not been matched by the realities of maintaining wilderness values. Although Romantic and aesthetic visions of wilderness have been important for valuing Australia's wilderness heritage, the reality is that utilitarian conservation has been paramount. Given the relative environmental resilience of wilderness areas under conditions of environmental change, as compared to that of disturbed areas, it is likely that a new series of wilderness values are about to be enacted, particularly as Australia grows increasingly concerned over environmental resource and water security. Such a situation will, therefore, most likely lead to a new series of contestations over the need to retain wilderness as part of the things that Australia wants to keep.

References

Adams, M. (2004) 'Negotiating Nature: Collaboration and Conflict between Aboriginal and Conservation Interests in New South Wales, Australia', *Australian Journal of Environmental Education* 20:1, 3–11.

Australian Tourist Commission (ATC) (2001a) *Brand Australia* (Sydney: ATC).

—— (2001b), *Corporate Plan 2001/2002–2005/2006* (Sydney: ATC)

Bardwell, S. (1979) 'National Parks for All – A New South Wales Interlude', *Parkwatch* 118, 16–20.

Committee of Inquiry into the National Estate (1975) *Report of the Committee of Inquiry into the National Estate, Parliamentary Paper No. 195, 1974* (Canberra: The Government Printer of Australia).

Finney, C.M. (1984) *To Sail Beyond the Sunset: Natural History in Australia 1699–1829* (Adelaide: Rigby).

Goldstein, W. (1979) 'National Parks – Queensland', *Parks and Wildlife* 2:3–4, 130–40.

Gössling, S. and Hall, C.M. (eds.) (2006) *Tourism and Global Environmental Change* (London: Routledge).

Hall, C.M. (1992) *Wasteland to World Heritage: Preserving Australia's Wilderness* (Carlton: Melbourne University Press).

—— (2006) 'World Heritage, Tourism and Implementation: What Happens After Listing', in A. Fyall and A. Leask (eds.). *Managing World Heritage Sites* (Oxford: Butterworth Heinemann).

—— (2007) *Introduction to Tourism in Australia*, 5th edn. (Melbourne: Pearson Education)

Hall, C.M. and Boyd, S. (eds.) (2005) *Nature-based Tourism in Peripheral Areas: Development or Disaster?* (Clevedon: Channel View Publications).

Hall, C.M. and Härkönen, T. (eds.) (2006) *Lake Tourism: An Integrated Approach to Lacustrine Tourism Systems* (Clevedon: Channel View Publications).

Hall, C.M. and Higham, J. (eds.) (2004) *Tourism, Recreation and Climate Change* (Clevedon: Channel View Publications).

Hall, C.M. and Page, S. (2006) *The Geography of Tourism and Recreation*, 3rd edn, (London: Routledge).

Helburn, N. (1977) 'The Wilderness Continuum', *Professional Geographer* 29, 337–47.

Herath, G. (2002) 'The Economics and Politics of Wilderness Conservation in Australia,' *Society & Natural Resources* 15:2, 147–59.

House of Representatives Standing Committee on Environment and Conservation (1981) *Second Report: Environmental Protection: Adequacy of Legislative and Administrative Arrangements* (Canberra: AGPS).

Kirkpatrick, J.B. (2001) 'Ecotourism, Local and Indigenous People, and the Conservation of the Tasmanian Wilderness World Heritage Area', *Journal of the Royal Society of New Zealand* 31:4, 819–29.

Kirkpatrick, J.B. and Haney, R.A. (1980) 'The Quantification of Developmental Wilderness Loss: The Case of Forestry in Tasmania', *Search* 11:10, 331–5.

Knight, R.L. (1999) 'Private Lands: The Neglected Geography', *Conservation Biology* 13: 223–4.

Lesslie, R. (1991) 'Wilderness Survey and Evaluation in Australia', *Australian Geographer* 22, 35–43.

Lesslie, R. and Maslen, M. (1995) *National Wilderness Inventory: Handbook of Prodecures, Content and Usage* (Canberra: Australian Heritage Commission).

Lesslie, R.G. and Taylor, S.G. (1983) *Wilderness in South Australia: An Inventory of the State's Relatively High Quality Wilderness Areas* (Adelaide: Centre for Environmental Studies, University of Adelaide).

Lesslie, R.G. and Taylor, S.G. (1985) 'The Wilderness Continuum Concept and its Implications for Australian Wilderness Preservation Policy', *Biological Conservation* 32, 309–33.

Marshall, A.J. (ed.) (1968) *The Great Extermination: A Guide to Anglo-Australian Cupidity, Wickedness and Waste* (London: Panther Books).

Mendel, L.C. (2002) 'The Consequences for Wilderness Conservation in the Development of the National Park System in Tasmania, Australia', *Australian Geographical Studies* 40:1, 71–83.

Mendel, L.C. and Kirkpatrick, J.B. (2002) 'Historical Progress of Biodiversity Conservation in the Protected-area System of Tasmania, Australia', *Conservation Biology*, 16:6, 1520–1529.

Mosley, J.G. (1978) 'A History of the Wilderness Reserve Idea in Australia', in J.G. Mosley (ed.). *Australia's Wilderness: Conservation Progress and Plans, Proceedings of the First National Wilderness Conference, Australian Academy of Science, Canberra, 21–23 October 1977* (Hawthorn: Australian Conservation Foundation).

Mulligan, M. (2001) 'Re-enchanting Conservation Work: Reflecting on the Australian Experience', *Environmental Values* 10:1, 19–35.

Nash, R. (1967) *Wilderness and the American Mind*, 1st ed. (New Haven: Yale University Press).

National Parks and Primitive Areas Council (1934) *Blue Mountains National Park Special Supplement, Katoomba Daily*, 24 August.

Oelschlaeger, M. (1991) *The Idea of Wilderness: From Prehistory to the Age of Ecology* (New Haven: Yale University Press).

Pettigrew, C. and Lyons, M. (1979) 'Royal National Park – a History', *Parks and Wildlife* 2:3–4, 15–30.

Pouliquen-Young, O. (1997) 'Evolution of the System of Protected Areas in Western Australia', *Environmental Conservation* 24, 168–81.

Powell, J.M. (1976) *Conservation and Resource Management in Australia 1788–1914, Guardians, Improvers and Profit: an Introductory Survey* (Melbourne: Oxford University Press).

Runte, A. (1979) *National Parks The American Experience*, 1st ed. (Lincoln: University of Nebraska Press).

Runte, A. (1983) 'Reply to Sellars', *Journal of Forest History* 27:3, 135-41.

Russell, J. and Jambrechina, M. (2002) 'Wilderness and Cultural Landscapes: Shifting Management Emphases in the Tasmanian Wilderness World Heritage Area', *Australian Geographer* 33:2, 125–39.

Sattler, P. and Creighton, C. (2002) *Australia's Terrestial Biodiversity Assessment.* Canberra: National Land and Water Resources Audit.

Seddon, G. and Davis, M. (eds) (1976) *Man and Landscape in Australia, Towards an Ecological Vision* (Canberra: AGPS).

Slade, B. (1985–86) 'Royal National Park: The people in a people's park', *Geo: Australia's Geographical Magazine* 7:4, 64–77.

Strom, A.A. (1969) 'New South Wales', in L. Webb, D. Whitelock and J. Le Gay Brereton (eds). *The Last of Lands* (Milton: The Jacaranda Press).

Syder, J. and Beder, S. (2006) 'The Right Way to Go? Earth Sanctuaries and Market-based Conservation', *Capitalism Nature Socialism* 17:1, 83–98.

Thompson, P. (1985) 'Dunphy and Muir – Two Mountain Men', *Habitat* 13:2, 26–7.

Tourism Australia (2006a) *A Uniquely Australian Invitation: Strategy & Execution* (Belconnen: Tourism Australia).

—— (2006b) *A Uniquely Australian Invitation: The Experience Seeker* (Belconnen: Tourism Australia).

Tuan, Yi-Fu (1974) *Topophilia: A Study of Environmental Perception, Attitudes, and Values* (Englewood Cliffs: Prentice Hall).

Waitt, G. (1997) 'Selling Paradise and Adventure: Representations of Landscape in Tourist Advertising', *Australian Geographical Studies* 35:1, 47–60.

Waitt, G. and Head, L. (2002) 'Postcards and Frontier Mythologies: Sustaining "Views" of the Kimberley as "Timeless"', *Environment and Planning D* 20, 319–44.

Waitt, G., Lane, R. and Head, L. (2003) 'The Boundaries of Nature Tourism', *Annals of Tourism Research* 30:1, 523–45.

Wright, J. (1977) *The Coral Battleground* (Melbourne: Nelson).

Chapter 5

Aborigines, Bureaucrats and Cyclones: The ABC of Running an Innovative Heritage Tourism Operation

Marion Hercock

Introduction

In this chapter you will be taken on an outback tour by the director of a small tourism enterprise. But don't expect a holiday, as this case study will demonstrate the complexities faced by a small business company which specialises in heritage tourism in the arid interior of Western Australia. Much of the larger scale subject matter and many of the issues raised in the other chapters resonate in this chapter at the micro-scale. These include issues of heritage protection, wilderness, heritage 'icons', indigenous places, migrant Australians, culture and places, and also sustainability.

Throughout the case study the author provides examples from the company's database and experiences.[1] From this evidence the reader can determine how little or much the small-scale reality differs from or reflects the general literature and wider context.

The case study is set out in four parts. The first part introduces a small business enterprise, and that company's use of heritage as a product appealing to a particular niche market (see Figure 5.1).

In the second part the company is briefly set into a wider economic and social context of global finance and markets. The third part of the case study examines the paradox of place and places, with particular attention given to the difference between *experience* and *place*; and the problem of marketing little known places. Finally, the fourth part covers, in depth, some of the 'on the ground' complexities created by the local social setting and the physical environment for the operator when running a remote area expedition. The social aspects include the government administration of conservation reserves; the private management of mining and pastoral leases and the administration of Aboriginal lands. These social aspects relate to the Western Australian state administrative and public policy regime in which the company operates. Allied to the social aspects, but also arguably physical in nature, are human impacts on the bio-physical environment. In conclusion, the prospects for the sustainability of the company (as a small business) and this type of heritage tourism are summarised.

---

1   Unless a specific source of data or information is cited, the authority is the owner/ operators of Explorer Tours, Marion Hercock and Jeremy Bryant.

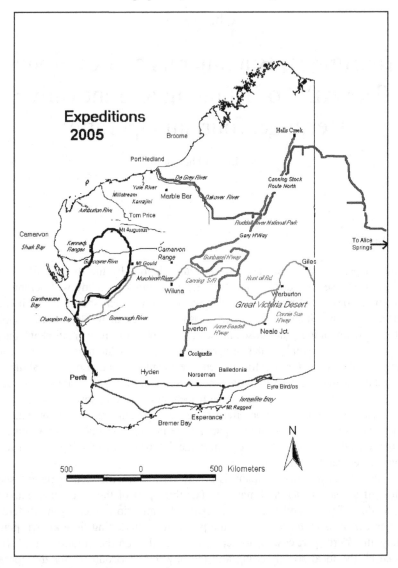

Figure 5.1   Map of tours offered by the company in 2005, indicative only

The Heritage Tourism enterprise

This case study is based on the experiences of a small tourism enterprise which consists of two owner-operators who manage the business and run the tours. Because the company does not employ any staff and has a turnover of less than $100,000 per annum it can be described as a 'micro-enterprise'. We started as a business entity in July 2000 and began running tours in March 2001 after industry research and field reconnaissance. Although the company operates within the niche market of heritage tourism, the product it offers also falls within the bounds of educational tourism, nature-based tourism, and soft-adventure tourism.

The Company's Heritage Tourism product

The company takes small groups of up to eight passengers on outback tours following the original routes taken by the nineteenth century explorers of Western Australia (see Figure 5.2).

Figure 5.2   Back page of Explorer Tours brochure 2004/5

This innovative approach is not entirely original, as tour companies elsewhere also offer tours retracing historical routes. For example, in Australia, North Queensland Wilderness Treks offer a tour retracing Edmund Kennedy's journey of 1848 (see www. wildex.com.au). Overseas, UK-based Peter Sommer Travels offer an archaeological tour 'In the footsteps of Alexander the Great' (www.petersommer.com); and Wild

Frontiers invites customers to 'join us in the footsteps of Genghis Khan' on a tour to Mongolia (www.wildfrontiers.co.uk). In all these examples, the products are being offered by well-educated, mature operators to a niche market.

In the case of Explorer Tours, each tour route is a themed journey or 'pathway' on which tour participants visit sites described by, named, or documented by the explorers, and/or where specimens were collected. Given the distances involved (from 3,000 to 8,600 kilometres) and the slow rate of travel through rough terrain, the duration of the trips ranges from nine to 26 days. A maximum of six scheduled one-off tours are offered in a year, with charter options. Tour parties visit and also camp at selected camp sites originally used by the explorers. For example, F.T. Gregory's Northwest Australian Exploring Expedition of 1861 formed the basis of one tour from Perth to the Pilbara and Gascoyne regions of Western Australia (Gregory 1884 [1968]). The exploration route provides a particular method and rationale for providing the tourist with a wide range of historical, environmental, economic, social and cultural interpretation. Original maps, journals and the published works by an explorer are the basis on which most of the tour route and its attractions and destinations are based. Following an historic exploration route is a heritage tourism experience in which the participants retrace the events of that original journey, revisit the discoveries made by the explorers and locate the legacies left by them in the course of their travelling holiday. On some tours, older Australians with a pastoral, mining or remote area medical background are happy to share their life experiences and knowledge with other tour participants. As a result, the value, topicality and relevance of heritage is enhanced and personalised.

The attractions and destinations on a tour include scenic landscapes, archaeological sites, pastoral properties and a range of heritage places, all of which are located in remote areas and linked by a particular exploration route. The attractions, together with the provision of expert guidance, camping equipment, outback dining, and transport through remote and often difficult-to-traverse terrain, make up the company's tourism product. The nature of the product and the interests of the customers ally this type of tourism experience to several other niches: rural tourism, natural heritage tourism, cultural heritage tourism, and gourmet tourism. These niches are not exclusive, but overlap in terms of places visited, activities and market. For example, some customers on a tour want to see sheep and cattle stations and stockmen at work. A larger sub-set of customers joins to experience isolation and wilderness, especially the desert, and to see stars and to camp in the outback. Others want to watch birds or photograph wildflowers. Some customers are more interested in Australian history and cultural heritage, which includes the past cultures of the indigenous nomads and the European settlers. Within this last group of customers, there is a sub-set who are more interested in Aboriginal culture and art, and meeting Aboriginal people. This information about customers' interests is collected from their application form to join a tour. Curiously, at least half of the applicants simply state 'being there [outback]' or 'out there' as their interest. This expression points to the important of sensory experience rather than that of visiting places, and will be discussed later.

Less tangible than built heritage, natural heritage, and cultural heritage such as art works, are the experiences associated with remote area travel in Australia. Many of these experiences, that are uniquely Australian, integrate Aboriginal, explorer and settler legacies. One example is 'bushcraft' or 'bushmanship', which is about adaptation to local conditions, interpreting the landscape and its resources for survival, and how best to move through that landscape, as well as understanding the night sky.[2] Learning about bushcraft in different habitats and regions can combine traditional local Aboriginal knowledge with the landscape reading skills of a geographer or environmental scientist. The successful nineteenth century explorers and prospectors mastered bushcraft, which was taken up by pastoralists, field naturalists and geologists. To the non-Australian, bushcraft is sometimes perceived as part of the Australian mystique; as demonstrated by the Mick Dundee character in the film *Crocodile Dundee* (Fairman, 1986). To many urban Australians, bushcraft is seen as part of their heritage. Bushcraft is not always understood or appreciated, but nevertheless, it is seen as 'Australian'. Travelling through the outback with, or meeting a skilled practitioner of bushcraft; or learning some of the basic skills, is an aspect of Australian heritage that tourists do not always see as heritage (as such), but is something that they associate with Australian culture.

Likewise, outback cooking, (that is, cooking on an open fire, often with a heavy cast iron camp oven), is something that is perceived as Australian. As with bushcraft, indigenous skills combine with European methods to produce something uniquely Australian. But, this cultural heritage is also on the wane, as a result of overuse of scarce firewood resources by increasing numbers of tourists. Sitting by a fire, or enjoying a meal cooked on a fire is an outback experience that is expected by outback travellers, and often taken for granted (see Chapter 6). The company promotes open fire cooking as part of the heritage experience it offers, with meals such as 'Windich's choice' and the 'Afghan cameleers' lunch' providing sustenance, and a means to educate customers about heritage and different foods.[3] However, the operators do emphasise to tour participants that the camp fire is something that *they* can enjoy, but that their grandchildren will probably never experience. The campfire is as an aspect of heritage that is unsustainable as the numbers of independent tourists to remote areas is increasing with the rising ownership of four-wheel drive vehicles (4WD). ·

A niche market

The company's use of heritage appeals to a particular niche in the tourism marketplace, which has been recognised and described by the Western Australian Tourism Commission (Tourism Western Australia – TWA).[4] For example, among local West Australian tourists, it is typically older travellers (aged over 55 years) who prefer their holiday experiences to be based on either 'places or interest' or

---

2 For a detailed analysis of the development of bushcraft in Australian exploration see McLaren (1996).

3 Tommy Windich (c.1840–1876), policeman and explorer. A Nyoongar from the York district in southwestern Australia, Windich was John Forrest's companion on several expeditions of exploration (Love 2003).

4 The Tourism Commission (Tourism WA) is a government agency, which is sometimes confused with the industry body, the Tourism Council.

'reconnecting'. Places of interest include historical sites, unique landscapes, nature and outback towns. TWA market research showed that for some older travellers experiencing history, heritage, rugged landscapes and nature is more than just sight seeing, it is also a 'reconnection' with the past and/or nature through learning. For example, some of the company's customers are revisiting country in which they have worked during their youth, and now, for a variety of reasons, they cannot, or do not want to enter the interior on their own. The favoured destinations include any place of historical or natural interest in Australia (TWA 2004a, 92).

By the close of 2005 the company had guided a total of 147 paying customers (about 50 per cent of whom are repeat customers) over five years of operation. Although this figure is not statistically valid, owing to the small size of the sample, some parallels with the official data are apparent. The company's customers are people who are predominantly older (50–80 years), better educated, often professional, local West Australians; some Australians from other states, and a few overseas travellers from Anglophone countries. Those customers tend to have some of the characteristics described in the literature on nature-based and heritage tourism (Weiler and Hall 1992). The overseas travellers have travelled to Australia on numerous occasions before, often to visit family. Having seen the popular tourist destinations, they are now seeking to spend more time in one area, to travel at a slower pace and to learn more about Australia. Of the total of eight overseas customers in the period 2001 to 2005, most conformed to the visitor profile identified by the WA Tourism Commission. For example, in 2003, UK travellers made up the highest number of visitors to WA, with a median stay of around 13 nights (TWA 2004b).

Market research by the West Australian Tourism Commission indicates that arts/heritage activities were undertaken by only 14.4 per cent of domestic West Australian tourists, with 37.5 per cent been involved in outdoor/ecotourism activities (TWA 2004a, 45). The national Australian figures are similar: arts/heritage 12.46 per cent and outdoor/ecotourism 33.94 per cent (TWA 2004a, 46). The researchers did not regard arts/heritage and outdoor/ecotourism as mutually exclusive activities, but activities which could be enjoyed together. These figures point to the small size of the market at which the company offers its tourism product.

The global and local setting of the heritage tourism company

While the company has been described a 'micro-enterprise' which supports the two owner-operators, it is, nevertheless, part of a much wider network of social exchanges: economic, commercial and educational. This network operates at the local level but also extends to the wider world through communications and finance. The national Australian and local West Australian administrative and public policy regime creates another sphere in which the firm must operate as a law-abiding and responsible entity.

The global economy and environment

In terms of the global world, the firm is physically *isolated* owing to its locational base in Perth, the State capital of Western Australia, but it is not *insulated* from world movements in finance and market trends. One obvious and immediate impact felt by the company is a rise in fuel prices. Diesel fuel accounts for 30 per cent of the cost of running a tour. In all, the expense of running the vehicles (fuel, maintenance and depreciation), is 60 per cent of the total cost of a tour. A rise in world oil prices in the middle of 2005 impacted heavily on the firm when local fuel prices increased.

The question arises – why not charge tour participants extra in response to fuel price rises? There are at least two reasons why an extra charge is not levied on customers. First, a year's programme of tours and prices has to be set at least six to nine months in advance of the forthcoming year, as potential customers need time to plan their holidays. While the younger section of the market tends to take shorter holidays with shorter lead times, the older section is freer to take longer holidays, which require more advance planning. Second, an immediate response could be to put a fuel surcharge on the set price of a tour. However, any surcharge would not be in the interests of good relations between the company and the customers, particularly in the case of a very small business that is dependant on repeat custom. In the main, people feel that once they have paid the cost of the tour, they should not have to pay any more than the specified amount.

Another influence on the company and its customers is the stock market and international money markets. The company's target market is over 50 years of age, professional, better educated or more widely read, and well-travelled people in Western Australia, Australia and in English speaking countries. Because a large proportion of this market consists of self-funded retirees who live on investments and/or their superannuation, this group is vulnerable to movement in the share market. Any slump in the share market or fluctuation in superannuation payments affects their personal budgets and purchasing patterns. For example, several of the company's local customers are paid share dividends by Wesfarmers.[5] When the dividends are increased, the recipients are more likely to spend. However, while a decrease in share dividends can result in customers not purchasing tours, a major increase in dividends does not necessarily result in more local customers for the company. A boom in share dividends, such as that experienced during the 2005/2006 'boom', may mean that local WA customers will spend on overseas trips rather than local tours. Such a move away from the domestic market might have resulted in a local slump. This movement was recognised as a problem by the Tourism Council of Australia in April 2006, when the figures released by Tourism Western Australia for December 2005 were published. These figures showed that Western Australia followed the national trend of a continuing decline in domestic visitors, domestic tourism expenditure and intrastate travel. The drop was 8.2 per cent in domestic

---

5 Wesfarmers Limited is a major Australian corporation, based in Perth. Founded in 1914 as a farmers' cooperative, the company was publicly listed in 1984 and is involved in oil and gas production and distribution, insurance, buildings products, agricultural chemicals and freight (Wesfarmers 2006).

visitors and 9.5 per cent in domestic tourism expenditure. The analysts cited a 24 per cent increase in the number of West Australians travelling overseas during 2005/2006 as one of the reasons for a decline in domestic tourism (TWA 2006). For the company, this pattern was reflected in responses to advertisements and places sold on the 2006 tour programme. Not one advertisement run in 2006 yielded a single inquiry, let alone an order. Excluding a charter tour for two overseas customers, six scheduled tours with a maximum of eight passengers gives a potential total of 48 customers, but only 29 customers bought a place on a tour.

Thus, while fluctuations in the share market and superannuation funds impact upon the company, the effects cannot be anticipated to be simple reflections of rises and falls. In brief, when the sharemarket is down, so too are local tour purchases, but when the sharemarket is up, local tour purchases may be either up or down.

We are often asked if terrorism, natural disasters and diseases such as SARS and bird flu affect the company and the purchasing pattern of its customers. Logically, it would seem that people might be attracted to travel in a country that is perceived as safe and clean, but the company has no evidence that those types of global phenomena have resulted in people purchasing our tours. The reader might look to the characteristics of the company's particular market for an answer. I will only offer the suggestion that an act of terrorism, a natural disaster or a pandemic might need to have a severe and sustained impact on the global stock market to affect the company's customers. In summary, the case study company is sensitive to, but has been resilient to minor or short-term fluctuations in the share market and fuel prices.

The Australian and local setting – government and administration

The company as a unit within a local economic network and a unit within the tourism sector, not only earns an income for its operators, it also adds to the income of other businesses in the tourism industry and other economic sectors. It is also a social unit which takes mainly middle-class urban dwelling people to meet people of different socio-demographic groups in remote areas. In this activity the company contributes to bridging the urban-rural divide. Although this latter aspect of the company warrants further discussion about social sustainability issues in remote areas, it is not discussed here. Instead, attention is given to a different division, that between administrative bureaucracies and private enterprise.

The operators accept that some government regulation of the company's financial activities, its conduct on public lands, and its vehicles and their operation, is reasonable in a liberal democracy. However, the cost to the small business does not solely affect the business, as the costs and the degree of regulation have implications for the longer-term protection of heritage. For the case study company, the costs of regulation include the following: company registration and business name, special licensing for drivers and vehicles, insurance, trademark, maintenance of equipment to required standards, and industry accreditation. In 2006, regulation cost the company $6,000 and that cost was recouped through income. Therefore, in order for the company to survive, it must make sufficient profit to meet the cost of regulation, normal operation and maintenance, and provide a useful income for the owner/operators.

However, instead of protecting consumers and heritage sites, regulation has the potential to lead to the very outcomes that it is designed to prevent. We know of 'cowboy' operators who avoid the cost of some licensing fees as the licence offers no gain (in customers), only a financial loss. Likewise, independent 'self-reliant' tourists are free to travel on public lands without identification and incur no penalty for transgressions, because they and their activities are unknown to the authorities. There is no financial or material benefit (privileged scientific and heritage information, access to camp sites) for the operator to have a licence to enter national parks. The fine for non-compliance is less than the cost of meeting the requirements for tour operators to enter national parks made by the state Department of Conservation and Land Management (CALM).[6] Bureaucratic antagonism towards commercial operations, exemplified by comments made by administrators, such as 'you are out to make a profit', and 'we know who you are and where you are going', are mildly offensive and do little for public relations.

Over-regulation of compliant, transparent businesses could lead to more unlicensed operators, and as a result, damage to heritage sites and the interests of consumers. For example, the CALM's Policy Statement on the 'Identification and Management of Wilderness and Surrounding Areas' expressly forbids commercial recreation and tourism operations within wilderness areas, while permitting 'self-reliant' [independent] recreational and educational expeditions (CALM 2003, s.4.10; 2006, s.4.10). Many heritage sites visited by the company are situated in national parks and nature reserves, such as Kennedy Range National Park (141,660 ha), which can be defined as wilderness (see Chapter 4). Yet it is responsible tour operators who educate their customers about the heritage of a place and control tour participants' behaviour and limit negative impacts, who are banned in favour of anonymous private individuals and not for profit groups.

Park managers are not the only authorities to penalise compliant companies for the actions of independent tourists. For example, on one occasion, access to a scheduled attraction on the John Forrest 1874 tour (the explorer's base camp on Aboriginal community land), was denied by the Ngaanyatjarra Council, a few weeks before the planned visit. Another tour operator later advised the company that this site had been closed because some stones had been removed from it. An inappropriate and irresponsible act by persons unknown had resulted in the closure of the site to persons known.[7]

In summary, the sustainability of heritage tourism is not aided by some regulations that give nothing of benefit to, yet mitigate against, law abiding tour operators, while unintentionally rewarding unlawful behaviour. A more inclusive approach between

---

6   This issue is discussed in more detail under the heading *Access to heritage sites: land managers permitting.*

7   Applicants for a permit to enter Ngaanyatjarra lands are required to submit the names and addresses of all people travelling, the make and registration of the vehicles, the proposed route, the purpose of the travel and the dates of entry and exit. The maximum time allowable in the lands is three days.

government regulators and the industry to controls on tour operators might, in the longer term, foster more interest in, education about, and better protection of heritage.

Marketing places – geography over all

The third part of the case study examines some of the paradoxes associated with *place* and places, which lead to a problem in marketing places. The people the company has guided on tours centred on Australian heritage fall into two main groups: the 'outback lovers' and the 'place collectors'. In the main, both groups do not have a primary interest in heritage as such (natural and/or cultural; tangible and/or intangible), but treat it as a bonus, rather than the sole reason for travel. The first group, the 'outback lovers', are those people who value the sensory experience of being in the outback, especially the semi-arid pastoral regions and the drier deserts, such as the Great Victoria Desert. The second group are far more interested in *places*. It is the places and destinations that are listed on an itinerary that attract these customers. The more places, the more attractive the itinerary is to the tourist, as it appears to offer more value for money. On tour, places are eagerly collected, in the form of photographs, journal entries or souvenirs, especially where the place is well publicised or has significant 'brag factor'.

Following early exploration routes means that unique and rarely seen places in remote areas are offered to the tourist. However, tourists cannot boast about visiting Warburton's Pillar or Central Mount Wedge in the Northern Territory because very few people know anything about these places (see Figure 5.3).

Figure 5.3   Central Mount Wedge, Northern Territory

Photograph by J.J. Bryant (2004).

In contrast, they can talk about their tour to El Questro cattle station or the Bungle Bungle Range in the Kimberley region of Western Australia, as these places are heavily promoted as 'icons', and are therefore well-known. The less sophisticated tourist can buy a souvenir at a major tourist destination, but teaspoons and tee shirts are not for sale in the wilderness.

Figure 5.4   Carnarvon Range, Western Australia

Photograph by J.J. Bryant (2003).

It is a paradox that in this market, people want to go to remote, unique places and do not want to encounter other tourists, but they will not go to a place if they have not heard or read about it. Also, once a beautiful, remote place becomes known, it is no longer remote and, as a consequence may have its beauty blemished or even destroyed. But without a public profile, little known, remote places do not sell. One example is the Carnarvon Range (see Figures 5.4 and 5.5) in the Little Sandy Desert.[8] Many West Australians are not aware of the existence of this place, confusing the range with the coastal town of Carnarvon (about 500 kilometres to the west) or the Carnarvon Range in Queensland, eastern Australia. Known to local pastoralists, exploration geologists, field naturalists and some life scientists and archaeologists, these sandstone ranges have all the features of a remote area heritage tourism site: past Aboriginal usage, links to exploration history, interesting landforms and biota, beautiful scenery and; at 30 kilometres distance, relatively close proximity to the widely known Canning Stock Route.

---

8   Named in 1874 by John Forrest, who sighted it in the distance, but did not visit.

Figure 5.5   Gallery at Serpents Glen, Carnarvon Range

Photograph by J.J. Bryant (2003).

Originally known as the Wiluna-Sturt Creek Stock Route when it was constructed over 1906–1910, the Canning Stock Route is globally known as a remote outback track for four-wheel drivers.[9] While the stock route has a high international and national profile and is of heritage significance, the nearby Carnarvon Range is barely known as a tourist destination. The company has received many enquiries about its particular tours that include a section of, or the entire Canning Stock Route. The stock route's profile may account for the popularity and success of our longest running tour (John Forrest's expedition of 1874, across the centre of Western Australia from west to east).[10] This 14-16 day tour, run each year in September, incorporates the water sources mapped by Forrest that were later exploited by the stock route construction team.

In summary, it appears that the economic viability of heritage tourism depends on the public profile of places. However, while that profile may sell places, it does not necessarily protect or maintain heritage. The problem of places being 'loved to death' might be reduced by government and industry promoting new destinations to the public eye, while reducing the profile of overly popular places. Any planning for such a reduction would have to consider those businesses which depend on tourism in a particular place. The closure of public access to some sites and places might not be possible or practical, but by greatly reducing the public profile, people are less likely to want to travel there.

---

9   An Internet search for the words 'Canning Stock Route' using any popular search engine will turn up hundreds of web references in a range of languages.

10   See Forrest (1875 [1998], 149–323).

Planning and running a tour

The fourth part of this chapter covers some of the 'on the ground' complexities created for the operator by the local social setting and the physical environment when running a remote area expedition. The social aspects include state government administration of conservation reserves and the private management of mining and pastoral leases, and Aboriginal lands.

The local social setting and the physical environment create unique challenges for the remote area operator as nothing is constant: rules change, permits once granted are denied, tracks are washed away, water sources dry up and vegetated areas are burnt out, fuel costs can soar. Both social and physical in nature (as they can damage sites and affect access to places) are negative human impacts, such as the creation of new tracks, the removal of timber, the removal of historical and/or cultural artefacts, graffiti and littering.

Physical aspects, which influence the conduct of tours, include the conditions of roads and tracks, prolonged drought and the occurrence of fire, extreme weather, rain and flooding all of which impact upon landscape aesthetics, the presence of wildlife and wildflowers and vehicular mobility. A prolonged drought, a fire or a flood can destroy an attraction while a severe rainfall event will render tracks impassable. There is a longer term question of climate change, as well as the cyclical effect of the El Nino- Southern Oscillation. In the six years that we have been operating in the interior, we have noticed variation in conditions over that time, and in comparison with the exploration records. This variation is interesting because historical writings often show a different distribution of water (reflecting the pattern and intensity of cyclones). For example, the water sources for the wells of the Canning Stock Route that were identified in 1906, follow palaeo-channels. In 1999–2000 many of these wells in the northern part of the Great Sandy Desert were flooded when the palaeo channels were recharged by rains from several massive cyclones. Examinations of past climatic patterns of the Australian monsoon might attempt to predict future patterns, but this long range forecasting has little application in the running of this business, given its short-term life. Nevertheless, such research provides useful material for interpretation and discussion on tour.

Access to heritage sites: land managers permitting

While discussed in greater detail in the next chapter (Chapter 6), the urban belief that the outback is a place in which Australians are free to roam is only a perception. Access to the outback and passage through land leased by pastoral companies and mining companies, and land owned by Aboriginal land trusts and communities, is not a free right. In order to pass through territory or to gain access to heritage sites, the company must first gain permission from land lessees and owners. While courtesy and common sense mean that the company must make itself and its intentions known before running a tour, the company always conducts a reconnaissance trip 12 months before planning a new route. Free of charge entry to wilderness is enabled by miners, pastoralists and some Aboriginal communities, who, each in their own way, reflect an aspect of Australian heritage. These groups often contribute towards a tour by providing travellers with information, or selling fuel, supplies and accommodation. In contrast, the state nature conservation agency (CALM) manages passage through, and access to heritage sites in national parks and nature reserves, by exercising more long-term control over the company itself, rather than its short-term actions. Tour operators must be licensed to take passengers into CALM land, and fulfil (and prove to have fulfilled) a range of legal, vehicular, insurance and operational requirements. Meeting those requirements costs the company $3,000 per annum. Annual registration fees are paid by tour operators to the agency, and additional charges are incurred every time a national park is visited.

Aspects of scale and cost are the primary differences between the government agency and the other land managers in their control of access to land. Mining and pastoral leaseholders and Aboriginal land owners control the company's actions only on the land for which they are responsible, and do not have any interest in the operation of the company. Their contact with the company is more immediate, short-term and seasonal. Since dealing with these groups often depends on the face to face development of mutual trust long before a tour is conducted, they know the operators as people, not just a business entity. The financial cost incurred by cultivating these relationships and gaining access to land is much less than that incurred when dealing with the government agency. CALM has more interest in the wider operation of the company and ensuring it meets its legal and vehicular requirements. In this regard, the department is repeating the regulatory work already carried out by the State Department of Transport (now the Ministry for Planning and Infrastructure) and the State Tourism Commission as well as the industry's own regulating body, the Tourism Council.

When arranging permits to travel through Aboriginal land, and to drive through or visit Aboriginal communities in Western Australia and the Northern Territory, the company faces particular complexities. One tour, run only in 2005, provides an example of the difficulties and uncertainties involved. The tour route followed that taken by Colonel Egerton-Warburton's expedition of 1873–1874, from Alice Springs in central Australia to the De Grey River on the north-western coast of Western Australia (Warburton 1875 [1981]). In addition to the permits for the tour route, permits were also required for the tour operators to drive the tour vehicles on the Great Central Highway (WA) and the Tjukururu Road (NT) to Alice Springs from Perth. The tour party needed eight permits from one state government department, two Aboriginal councils and three communities to travel through different sections of the proposed route and to visit sites. Each administrative bureaucracy has different requirements from permit applicants and different regulations, although general conditions, such as travel within three days of the approved date, apply. Likewise, the communities vary in their rules concerning transit and visits. The WA state government has the simplest requirements and most efficient processes. The councils require more personal detail about all the trip participants, and must first consult with the relevant communities and land trusts before considering applications at a general meeting of council members. This process take can take weeks so applications must be tendered well in advance of a tour. Approval from individual communities must be gained for excursions to sites which are not on the roads and tracks stated in the application. In some instances, even if the relevant council permits have been

Figure 5.6a   Wooleen Station shearing shed, built in 1922

Photograph by M. Hercock (2002).

Figure 5.6b   The shearing station in 2006, after destruction by storm in December 2004

Photograph by M. Hercock (2006).

granted, final clearances from the community for travel to a site have to be obtained a week before and again, a day before.[11]

Planning before any tour involves drawing up a range of alternative routes and stopping places, but uncertainty over access when a tour is actually underway makes the operation much more difficult. Refusal of permission to travel a section of a tour route can result in a diversion of several hundred kilometres or rerouting a major section of the tour. Thus for every plan, there an alternative route, plan B; and even a plan C. However, when a tour has to be rerouted without warning, and sites are omitted from the itinerary, for no logical reason, such as floods, it results in disappointed and unsatisfied customers.

Access to heritage sites: weather permitting

The most difficult to manage element in running tours to heritage sites in remote areas is the weather. Cyclones and floods render travel impossible on outback tracks and can result in damage to or the destruction of heritage sites. For example, soaks in deep sand, such as Patience Well in the Great Sandy Desert, once used by Aboriginal nomads to access water stored in underground channels, are silted up during a flood. Without regular clearance of sand and debris these sites disappear. In contrast, visits by tourists to Ngarinarri Soak, on the western margins of the Gibson Desert, have resulted in the preservation of the soak as some tour parties have been maintaining this native well. Ngarinarri Soak is the site where Warri and Yatungka, the Mandildjara couple known as 'the last of the nomads' were found in 1977 (see Peasley 2002). The few tour operators who visit the soak consider it to be an important heritage site of national significance.

Equally vulnerable to cyclonic winds and storms, are built heritage sites. The loss of a heritage site can have a social and economic impact beyond an immediate physical loss. One example is the destruction of the Wooleen Station shearing shed by a freak storm in December 2004 (Zekulich 2005, 39). The shed (see Figure 5.6), built in 1922 was heritage-listed and a working component of the sheep station, as well as being part of Wooleen's own 'station stay' tourism enterprise (Nixon and Lefroy c.1989, 198–199).

The Murchison Shire and the Midwest Regional Development Authority had promoted the shearing shed as a destination for heritage tourists as part of a campaign to diversify the local economy; and several Geraldton and Perth-based tour operators had included the shearing shed on tour itineraries.

Australian Bureau of Meteorology climatic data enable the company to programme an annual tour schedule to take advantage of optimal conditions for travel. Therefore tours to the semi-arid and arid interior are run in the driest, but coolest months between April and September. However, storms and rain can still occur. From October to March, summer cyclones (the northern Australian monsoon) and extreme temperatures (35° to 45°C) can render travel extremely difficult and

---

11 Applicants who question the permit system risk being banned from entry to Aboriginal lands. The control of access and the tightness of closure of communities are discussed by Rothwell (2006, 25), who echoes the experiences and observations of the company.

unpleasant. While short-term forecasts are useful in providing a guide to conditions before a tour departs, a final check has to be made of track conditions, especially in the event of a late season cyclone.

A storm event weeks in advance of a tour can result in inaccessible roads and tracks and flooded sites. For example, during 29–31 March 2006, Cyclone Glenda, a category 4 storm off the Pilbara coast, and tracking south-west towards the interior of the state, occurred two weeks before a charter tour and a month before a remote area tour. The cyclone was a threat to the scheduled tours because widespread flooding in the Ashburton, Gascoyne and Murchison River catchments would render regional roads and tracks impassable. These northern water courses are braided streams with catchment areas the size of England or Bulgaria, and rainfall many kilometres away from the main channels can still result in a rising river. An analogy would be rain in Edinburgh causing floods in London.

In summary, any rain or storm event in the interior creates uncertainty about tracks, mobility, and the state of attractions.

Summary

It is the experience of the company that it is *place* that matters, and the public perception of places determines the desire to visit. For a place to be a marketable destination, it must have some public profile or an image. Heritage contributes to making this image, and may be indivisible from the mystique of a place, but heritage as such, does not appear to sell places.

The company rates the success of an itinerary in terms of seats sold; but the success of a tour is determined on the ease of access to attractions, good weather, new discoveries, and the participation of interested, appreciative and involved passengers.

For the company, heritage tourism has provided an unparalleled way of life rather than a sustainable livelihood. Commercial viability in the longer term would depend on forming alliances with larger national and international touring companies, such as AAT Kings or World Expeditions. In the meantime, we will continue to drive and walk to some very remote and interesting places, learn more about our heritage, meet new and interesting people, and enjoy the sense experience of the outback while being paid to do so!

References

CALM (2003) Draft Policy Statement: Identification and Management of Wilderness and Surrounding Areas (Kensington: Department of Conservation and Land Management). Also available on the internet (reference below).

CALM (2006) Policy Statement 62 : Identification and Management of Wilderness and Surrounding Areas (Kensington: Department of Conservation and Land Management). Also available on the internet (reference below).

Fairman, P. (director) (1986) *Crocodile Dundee*, Rimfire Films.

Forrest, J. (1875 [1998]) *Explorations in Australia*, (London: Sampson Low, Marston, Low & Searle) [Facsimile edition, Adelaide: The Friends of the State Library of South Australia].

Gregory, F.T. (1884 [1968]) 'North-West Coast 1861' in A.C. Gregory and F.T. Gregory F.T., *Journals of Australian Explorations* [Facsimile edition, New York: Greenwood Press] 52–98.

Love, M. (2003) Museum News – Finley statues of Tommy Windich and Tommy Pierre, *History West* 41:1, 3-4.

McLaren, G. (1996) *Beyond Leichhardt: Bushcraft and the exploration of Australia* (Fremantle: Fremantle Arts Centre Press).

Nixon, M. and Lefroy, R.F.B. (c.1989) *Road to the Murchison: An illustrated story of the district and its people* (Murchison: Shire of Murchison).

Peasley, W.J. (2002) *The Last of the Nomads* (Fremantle: Fremantle Arts Centre Press).

Rothwell, N. (2006) 'Men's secret out', *The Weekend Australian* Weekend Inquirer, 27–28 May.

Tourism Western Australia (2004a) *Australia Market Profile* (Perth: Tourism Western Australia).

Tourism Western Australia (2004b) *United Kingdom Market Profile* (Perth: Tourism Western Australia).

Tourism Western Australia (2006) 'Domestic Market Quick Quotes, December 2005, Sourced from the National Visitor Survey', Tourism Research Australia, Research and Analysis (Perth: Tourism Western Australia).

Warburton, P. (1875 [1981]) *Journey Across the Western Interior of Australia* (London: Sampson Low, Marston, Low & Searle) [Facsimile edition, Victoria Park: Hesperian Press].

Weiler, B. and Hall, C.M. (1992) *Special Interest Tourism* (London: Belhaven Press).

Zekulich, M. (2005) 'Historic shearing shed hit by storm', *The West Australian*, 5 January 2005, p.39.

Internet-based references

Department of Conservation and Land Management [website], Managing Wilderness Areas (published 2003) http://www.naturebase.net (home page), accessed 30 June 2003.

Department of Conservation and Land Management [website], Managing Wilderness Areas (published 2006) http://www.naturebase.net/national_parks/management/wilderness_management.html , accessed 26 May 2006.

'Domestic tourism slump continues', *Tourism Talk, Tourism Industry News*, Weekly Round-up for Western Australia, on line industry magazine (published online 5 April 2006) http://www.tourism-talk.com.au (home page), accessed 7 April 2006.

North Queensland Wilderness Treks [website], http://www.wildex.com.au (home page), accessed 15 May 2006.

Peter Sommer Travels [website], http://www. petersommer.com (home page), accessed 23 May 2006.

Wesfarmers [website], http://www1.Wesfarmers.com.au (home page), accessed 25 May 2006.

Wild Frontiers [website], http:// www.wildfrontiers.co.uk (home page), accessed 23 May 2006.

Chapter 6

Waltzing the Heritage Icons: 'Swagmen', 'Squatters' and 'Troopers' at North West Cape and Ningaloo Reef

Roy Jones, Colin Ingram and Andrew Kingham

Introduction

Issues of contested (Shaw and Jones 1997) or dissonant (Tunbridge and Ashworth 1996) heritage are commonly associated with urban environments since these have traditionally been, as Barthes (1981, 96) observes, 'the place of our meeting with the other'. Certainly, over the last half-century or more, Australian cities have become increasingly culturally diverse. And, albeit more recently, and more slowly, at least some of these diverse heritages have gradually been acknowledged (Anderson 1999, Burnley 2005). However, with some localised exceptions, such as the Sikh community on the New South Wales North Coast at Woolgoolga, or the Italians in the Murrumbidgee irrigation district, non-Anglo Celtic (to use the common Australian term) migrants have tended to settle in and around the major state capitals, rather than in rural and remote areas. Indeed much non-Indigenous settlement in remote Australia has been both transient and exploitative and, even in the broad acre wheat and sheep farming areas, rural depopulation has been the norm in recent decades (Jones 2001).

For much of Australian colonial and postcolonial history, therefore, heritage and, indeed, more general, contestation, in rural and remote areas took place between Indigenous and predominantly Anglo-Celtic settler groups (Reynolds 1990). By the early twentieth century Indigenous dispossession was effectively complete, though belated recognition of Aboriginal occupance and heritage values in these regions was accorded by the Mabo and Wik decisions of the High Court in the 1990s. In recent decades, however, Australia, like most developed countries, has been experiencing a 'multifunctional rural transition' (Holmes 2006). While, as Holmes argues, it would be an oversimplification to see this transition as a straightforward shift from a productive to a postproductive socio-economic emphasis in many of Australia's rural and remote regions, a major component of this shift is undoubtedly the move from primary production to a wider and more complex range of rural activities with a concomitantly wider and more complex appreciation of the various heritage issues thus entailed. What has occurred in many rural areas in recent decades, therefore, is a shift towards 'contested countryside cultures' (Cloke and Little (eds) 1997) in which different sections of the largely settler, largely Anglo-Celtic population have 'othered' each other in rural and remote Australia.

According to Holmes (2002), the key factors driving this transition are: changing social values; agricultural overcapacity; and the rise of alternative, amenity-oriented rural land uses. As this transition occurs, as different people move in, as existing residents take on new economic and social roles and as urban dwellers begin to experience and to evaluate non-metropolitan Australia in new ways, so too do different aspects of the Australian rural and natural environment and, thus, of Australia's rural and natural heritage come to be perceived and valorised in novel and frequently differing forms. While such a categorisation is by no means comprehensive and, still less so, mutually exclusive, these perceptions of rural and remote Australia can be broadly classified in terms of the relative significance that they attach to the issues of production, consumption and protection.

Elsewhere in the national literature, considerable attention has been paid to the 'intense competition on values' (Holmes, 2006, 144) and to the planning and heritage preservation challenges that have therefore occurred in the scenic 'sea change' (Burnley and Murphy 2004) and 'tree change' areas in relative proximity to the major state capitals (Selwood et al. 1996; Tonts and Grieve 2002). In one of the few studies on the more remote regions of the country, Holmes (2002) has described recent changes in Australia's rangelands as 'a post-productivist transition with a difference?' (albeit with the concluding question mark) and, inevitably, the massive differences in scale, environment, and population size and density do set the rangelands apart from the more settled areas of Australia. Yet, perhaps because of their very isolation (until recently) and their sparse populations, rangeland areas have the potential to illustrate, in a particularly stark manner, the tensions between the values of production, consumption and protection.

This chapter describes how these three values, and the heritages with which they are imbricated, intersect in a particularly iconic 'outback' area, the North West Cape –Ningaloo Reef region, some 1,100–1,400 kilometres north of Western Australia's state capital, Perth (see Figure 6.1).

On land, the Cape Range National Park is a scenic and ecologically significant region, but its conservation importance is overshadowed by that of the Ningaloo coral reef, located a very short distance offshore. Protection of these natural resources is vital in scientific terms. Yet, not surprisingly, the local economy is now dominated by a rapidly growing (eco) tourist industry which is almost completely dependant on the natural endowments of the region (Carlsen and Wood 2003). As tourist numbers grow, protection values clash increasingly with the consumption values espoused by tourists, and particularly by those tourists who perceive it to be their birthright, as Australians, to roam (and camp and fish) in what they see as the 'wide open spaces' of their native land. However, this tourist industry is a very recent local phenomenon. The first settler activity in this region was pastoralism and, as has been the case for more than a century, sheep stations still occupy much of the North West Cape/Ningaloo region. While these stations could be said to represent the productive values of this remote area, their long-term economic viability is widely seen as being marginal at best. To further complicate the heritage picture, 'outback' pastoral

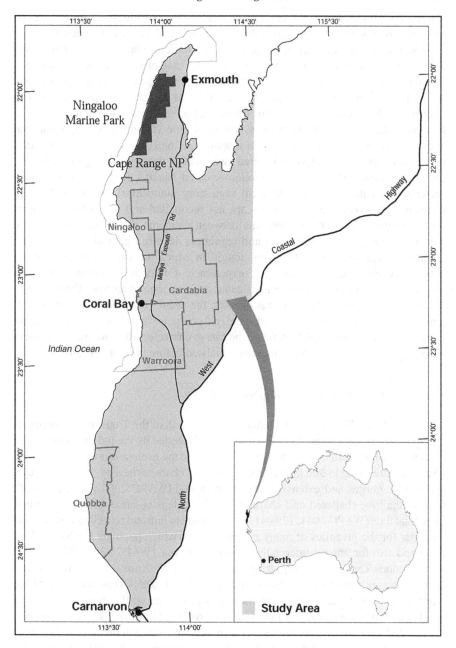

Figure 6.1   North West Cape-Ningaloo region

Source: Carlsen and Wood 2003, 23.

stations are also seen as representing a pioneering way of life which has long been valorised as an iconic component of Australia's national heritage and identity.

The intertwined heritages of North West Cape therefore involve a water body and a shoreline (of sufficient value, particularly in the case of its coral reef, for these to be environmentally managed and protected), campers in the outback and pastoralists on long-established sheep stations. In searching for a framework within which to describe the interactions between this concatenation of heritage icons, it is therefore hard to avoid the use of Australia's equally iconic (though unofficial) national anthem, 'Waltzing Matilda' where, in a waterside, outback setting, a contestation occurs between a consumption-oriented swagman (a camper making unauthorised use of the local resources), a production-oriented squatter (a pastoralist) and the protection-oriented troopers (the local regulatory authorities). In the ballad, the differences between the protagonists are not reconciled and the ending is tragic. At North West Cape, though tensions between elements of the local population, the tourists and the land managers and regulators have run high in recent years, a mediated outcome is currently being sought in which the varied heritages of the natural environment, of the 'laid back' experience of wilderness camping and of the lifestyle of the remote area pastoralists can remain as something more than ghosts.

In the following sections of this chapter, the scene (the metaphoric billabong, jumbuck and coolabah tree, in this case the real components of the North West Cape-Ningaloo environment) will be set, and the cast of characters (the swagmen, squatters and troopers) will be described and contextualised before the story is told.

The scene: Cape Range and Ningaloo Reef

Australia's North West Cape, immediately to the north of the Tropic of Capricorn, is the tip of a peninsula some 100 kilometres long bounded by the Indian Ocean to the West and the Exmouth Gulf to the East. The spine of the peninsula is the Cape Range, a karst feature rising to 300 metres above sea level characterised by 'narrow valleys, spectacular gorges and extensive cave formations' (WAPC 2004, 172). Exmouth Gulf is shallow, sheltered and characterised by extensive areas of mangrove and inter-tidal flats (WAPC 2004, 179) which contribute to nutrient recycling and provide a habitat for the juveniles of many marine species which populate the North West Shelf and also for 'the commercially important prawns' (WAPC 2004. 183).

Ningaloo Reef is the only fringing coral reef in Australia. It extends 260 kilometres south from the Murion Islands along the Indian Ocean shoreline of the North West Cape. Since the adjacent, dunal coast has been, until very recently, almost uninhabited, it is one of the most pristine coral reefs in the world. As a fringing reef, it is located very close to the shore, never more than a few kilometres and in places only tens of metres from the beach. As such, it is far more vulnerable to any negative environmental impacts of onshore development than, for example, the Great Barrier Reef. Climatically, this region is extremely arid, with the annual average evapotranspiration rate of 2, 591 mm. far exceeding the annual average rainfall of 226mm. The area is subject to tropical cyclonic activity between January and March and cyclones (notably Vance in 1999) and even a 1994 tsunami have caused damage to the reef in recent years. The valuable and complex local ecosystems are therefore vulnerable to natural as well as human-induced pressures.

While the marine and, to a lesser extent, the land resources were sufficient to support a small Indigenous population, the region appeared to offer limited productive opportunities for early settlers. Nevertheless, in the late nineteenth and early twentieth centuries, pastoralists gradually moved into the North West. Much of the (Crown) land adjoining Ningaloo Reef passed into pastoral leasehold tenure from 1876 onwards. Land transport connections with Perth were virtually non-existent in the early twentieth century, but the pastoralists built up sheep populations and exported their wool from local jetties, notably from the (now derelict) jetty at Maud's Landing.

Two lighthouses and a whaling station were established in the area shortly before World War One and the region was regularly visited by other whalers, and by pearling and fishing vessels. However, it was not until the area gained considerable strategic significance following the fall of Singapore to the Japanese in World War Two that further land-based development occurred. In 1942 the United States Navy established a submarine base on Exmouth Gulf, to the south of the present Exmouth townsite. The Australian support base, which offered radar, radio and fighter cover for the base eventually became the present RAAF Learmonth and the site for the region's (civil) airport. In 1962, the Australian and United States Governments signed an agreement for the construction of a Cold War Communications base at North West Cape. The town of Exmouth, the first urban settlement in the region, was constructed to support the base and the town and the base were both officially opened in 1967.

By the 1960s, both North West Cape and Ningaloo Reef were becoming appreciably less isolated. Commercial prawning and pearling operations were set up in or near Exmouth, communications improved with the sealing of the road from Perth as far as the regional centre of Carnarvon (400 kilometres to the south), and a massive increase in investment, development and population occurred in the adjoining Pilbara region in conjunction with an iron ore mining boom. Adjacent to Ningaloo Reef and 165 kilometres south of Exmouth, tourism development commenced with the construction of a hotel at Coral Bay. Further road sealing, to Exmouth and Coral Bay in the 1980s, and to the northern part of the reef in the 1990s, further encouraged access and tourism development. Indeed tourism growth enabled Exmouth to survive the massive downgrading of the military base at the end of the Cold War. In 1993, 700 Americans left this town of ca. 2,000 people, and their 185 houses were placed on the market. A five-year plan was envisaged for their disposal. But, given the rapid growth of the local tourism industry and their relatively low sales prices, they were all sold within one year.

Parallelling the growth of tourism has been a growth in conservation activity. A ca. 13,000 hectare reserve was gazetted in the Cape Range in 1964. Its status was raised to that of a National Park in 1971 and it was extended to an area of ca. 50,000 hectares through the inclusion of former pastoral lease land in 1974. Marine conservation occurred later. The Ningaloo Marine Park, which extends well beyond the reef and includes over 4,500 square kilometres of both State and Commonwealth waters as well as a 260 kilometre long and 40 metre wide coastal strip, was first proposed in 1974 and finally gazetted by both the Western Australian and Australian governments in 1987. By the late twentieth century, therefore, the stage had been set for the interaction of production, consumption and protection interests at Ningaloo and North West Cape.

The cast of characters

Swagmen/wilderness campers

Over 200,000 tourists visit the Gascoyne region annually, with over 130,000 visiting Exmouth alone. The majority of these are relatively short stay and, frequently international/interstate tourists using hotels, motels, backpacker accommodation, caravan parks or officially designated campgrounds. By contrast, wilderness campers at Ningaloo select undesignated coastal campsites in remote parts of the pastoral stations, or on the Commonwealth Department of Defence-owned coastal strip of land between the Cape Range National Park and the northernmost of the pastoral stations. Their numbers are smaller and they are spread out over some 200 kilometres of coastline albeit tending towards clusters at popular nodes. Wilderness camper numbers have increased rapidly in recent years. On Ningaloo and Waroora stations, the number of campsites grew from 131 in 1995 to 318 in 2002 (Armstrong 2003, 17). Furthermore, wilderness campers tend to stay for far longer periods than do other types of tourists. In 2001 and 2002, their average length of stay was 47 days (Remote Research 2002).

Figure 6.2   'Wilderness' camp site

In demographic terms, these are a predominantly Western Australian, and, on balance, an elderly group who are passionately committed to this form of recreation. The vast majority, almost 90 per cent in surveys conducted in 2001 and 2002 (Remote Research 2002), are from intrastate, and are mainly from Perth which contains over two thirds of Western Australia's population; 87 per cent in 2001 and 77 per cent in 2002 were revisiting the area and, in the 2001 survey which was conducted during school term time, over half the respondents were seniors. They valued the isolation and the scenic beauty of the area and frequently had strong attachments to particular campsites. Indeed some mark their 'patch' for the following year and/or work with the pastoralists on local revegetation and other environmental restoration projects. Even though they characteristically required high value camping, boating and 4-wheel drive equipment for this type of experience (Figure 6.2), they appreciated the low charges levied by the pastoralists for providing them with access to this 'wilderness experience'.

Anecdotal evidence would indicate that a controversial extension of sanctuary (i.e. non-fishing) zones from 10 per cent to 34 per cent of the marine park in 2005 may have deterred some Western Australian campers. On the other hand, Ningaloo's fame is growing nationally, with a number of interstate campers extolling this stretch of coastline as the 'last place' in Australia where this type of recreational experience could be found on the popular 'Four Corners' television programme, broadcast in August 2006. It is therefore likely that the camper pressure will continue to build along this coastline.

By bringing their 'homes' and many of their supplies with them – a large number are so-called silver nomads from the temperate south seeing out the winter in the tropics – and by obtaining a proportion of their food from the ocean, they attain a degree of self-sufficiency. However, one aspect of these traits is that they contribute substantially less (at least per day) to the local economy than do other types of tourist – though they do tend to stay in the region for considerably longer. Furthermore, notwithstanding the high environmental ideals espoused by many and practised, to a greater or lesser extent, by some of this group, they have at least the potential to generate severe impacts on what is an extremely fragile and valuable ecosystem unless greater levels of stewardship and environmental management can be implemented.

Even though this is now an unacceptable and a minority activity, some wilderness campers still arrive with trailers and freezer equipment and undoubtedly fish for significantly more than their immediate, camping holiday needs. In July 2001, 79 per cent of campers reported using campfires. Even though this proportion dropped significantly in the 2002 survey and local campfire bans have been proclaimed, it is difficult to enforce such regulations. In a near desert environment, the negative impact of this on the local vegetation and dune systems is potentially severe. In both 2001 and 2002 only a small proportion of respondents used the environmentally preferred 'long drop' toilets and the vast majority either buried or burnt their rubbish or used local tips on the pastoral stations which were, on occasion, little more than inadequate depressions. Given the proximity of the campsites and local tips to the reef, the high average wind speeds and the considerable violence of the infrequent rainfall events, the potential for refuse material to be washed, blown or leached into the ocean and thus to contaminate the nearby reef is considerable.

Finally, the area is virtually inaccessible other than by four-wheel drive vehicles. In a sand dune environment, tracks can readily become impassable while others can, equally readily, be created. Local aerial photography clearly indicates that, as camper numbers have increased, some four wheel drivers have 'bush bashed' the vulnerable vegetation and dune systems creating complex and excessive patterns of new tracks. Even though some campers block off redundant tracks for revegetation, this is an ongoing problem given the dunal nature of the local environment.

In the relatively short space of a few decades, these reefside camping grounds have therefore become the site of a quintessentially Australian outdoor experience involving fishing, camping, snorkelling on a coral reef and four-wheel driving, and this has, for many of its adherents, become a regular seasonal activity. But, as the quantity of wilderness campers grows, the question of how the quality of their recreational experience can be sustained becomes an increasingly vexed one.

Squatters/Pastoralists

The story of the pastoral occupation of Australia is an epic one (Heathcote 1965). Between the early nineteenth and early twentieth centuries, vast areas of Australia were occupied by sheep and cattle herders as they moved to progressively more remote and environmentally challenging areas. Initially this movement was unauthorised and the early pastoralists were, indeed, squatters on land that the colonial authorities had claimed but did not have the resources to control. In order to regularise this situation, the colonial governments devised leasehold systems whereby the land remained Crown property, but the occupiers were given the security of relatively long-term leases, provided that they conformed to a range of conditions relating to pastoral land use. Inevitably the 'tyranny of distance' (Blainey 1966) meant that government control over the more remote properties was limited and that, for a time, many early pastoralists were indeed 'Kings in Grass Castles', to quote the title of Durack's (1959) family memoir.

This romantic ideal of the pioneering, independent pastoralist struggling in a harsh, often scenic, but always stereotypically Australian, environment is an enduring component of the Australian legend. It has, almost from its inception, been famed in song (not only 'Waltzing Matilda') and story, ranging from Aeneas Gunn's 'We of the Never-Never' (1908), through Xavier Herbert's 'Capricornia' (1938) and Nevil Shute's 'A Town like Alice' (1950) to Baz Luhrmann's Nicole Kidman-Hugh Jackman epic, (provisionally and revealingly titled 'Australia') which is due to start filming in the Kimberley in the near future. Indeed, another Shute novel, 'Beyond the Black Stump' (1956), which describes the interactions between the owners of a Western Australian sheep station and an oil exploration team, could have been describing the events at North West Cape at that time. Certainly Shute was noted for the meticulous geographical research from which he crafted his novels (Jones 1996).

Over the course of twentieth century, however, the sustainability of the pastoral industry has been brought into question on a variety of grounds. The expansion of more intensive forms of agriculture and stock rearing gradually displaced pastoralists from any well-watered, fertile land with reasonable market access. And, indeed, the tendency of stock to rapidly graze out the most suitable plants had in any case caused many pastoralists to keep moving to more and more remote regions in search of fresh pastures (Bolton 1981). In the early to mid twentieth century, the economic viability of many pastoral stations was, in part, sustained by the use of an Aboriginal labour force paid in kind and/or at lower wage rates than those that applied elsewhere (see Gill and Paterson, in this volume). Constitutional and industrial changes in the 1960s that gave Aborigines the same entitlements as other employees therefore impacted severely on many marginally viable pastoral stations.

In the late twentieth century, the environmental and legal pressures on many pastoralists increased. Growing concerns for native species of flora and fauna and over the impact on vulnerable arid ecosystems of overstocking and of the spread of exotic and feral species have led to increasing restrictions being placed on pastoral activities and on more land (such as that now in the Cape Range National Park) being transferred from pastoral to conservation uses. The 1992 Mabo judgement enabled Indigenous groups to claim Native Title rights over crown land and the 1997 Wik case confirmed that, in certain circumstances, the granting of pastoral leases did not necessarily extinguish these rights.

As Holmes (2002) points out, pastoralists face an uncertain future across Australia's rangelands. In Western Australia the future horizon for all pastoralists is very much focussed on 2015. In that year, all pastoralists in the state must apply for the renewal of their leases and the Pastoral Lands Board has been developing a 'pastoral exclusion process' (Keys 2006, 12) which will identify an estimated one million hectares of land which will then be excluded from the pastoral leases for Aboriginal, recreational or conservation purposes. For the owners of the pastoral stations adjoining Ningaloo Reef, the Lands Board review is particularly relevant. Current proposals involve the removal of a two kilometre wide coastal strip from all stations as a general principle, with more extensive excisions in areas of particular recreational or conservational importance. For local pastoralists facing uncertain economic and environmental conditions, a 'post productive transition' into wilderness camping tourism has provided a valuable alternative income stream in recent years. The excision of the coastal campgrounds from their leases could well bring the overall viability of their operations into question.

Troopers/environmental managers

Administration of the pastoral leases was initially carried out by the Colonial and, from federation in1901, the State government in Perth though, at least in the early decades of pastoral settlement, their ability to oversee the extent to which the pastoralists conformed to their lease conditions was extremely limited. In the early twentieth century the establishment of a lighthouse indicated some desire for government oversight of marine activities in the area and, gradually, state government departments took a growing interest in Western Australia's fishing, prawning and pearling industries.

Figure 6.3   'Managed' camp site

Until recently, however, environmental controls over land-based activities in the North West Cape-Ningaloo area were limited. Most of the major developments in this region had been rationalised on the basis of military necessity and, as late as the 1960s, the local – though somewhat distant – Shire of Carnarvon had been unable to prevent a potentially damaging tourist development proceeding at Coral Bay. In the latter part of the twentieth century, this situation changed radically as a steadily growing area of land and sea was formally accorded conservation status between the 1960s and the 1980s and as the situation of the town and Shire of Exmouth became increasingly 'normalised' with the post Cold War winding down of the American military base.

While the designations and expansions of the Cape Range National Park and the Ningaloo Marine Park were made by governments in Perth and also, in the latter case, in Canberra, these decisions necessitated the build up of a local environmental management presence above and beyond that which had been provided by fisheries and pastoral inspectors. Over the last thirty years, therefore, the state Department of Environment and Conservation (formerly Conservation and Land Management or CALM) has become a not insignificant presence in the area. A district office has been established in Exmouth, a visitor and interpretation centre has been built adjacent to the reef, several (managed) camp sites have been established (Figure 6.3) and numerous direction and interpretation features have been provided throughout the Cape Range National Park. The Department also has a marine remit which requires it to monitor activities such as recreational fishing and human/marine animal interaction within the designated conservation areas.

What this has produced, however, is a somewhat bifurcated system of environmental management. Most of the land backing the northern third of the reef is in the Cape Range National Park (Figure 6.1). Here entry to the park is monitored, access and camping fees are charged, campsites have designated capacities, marine sanctuary zones can readily be checked for unauthorised fishing activity and tourist and tourist business operator behaviour can be directed and controlled. South of the Park and, more specifically, 'beyond the bitumen' (which currently extends from Exmouth through the Park to its southern boundary at Yardie Creek) the situation changes. Legally the Marine Park boundary extends 40 metres inland from the high water mark for the entire length of the reef, but from Yardie Creek southwards this boundary adjoins either Commonwealth Department of Defence land (sometimes used for firing range purposes) or crown land currently held as pastoral leases. The Department of Environment and Conservation has a draft agreement with the Department of Defence concerning the environmental management of this land. But, in practice, it is difficult for the Department (a state, rather than a commonwealth instrumentality) to exercise its environmental management responsibilities over two thirds of the marine park's shoreline. In addition to any legal/land ownership issues, much of this shoreline is over 100 kilometres from the nearest settlement and only accessible by dirt tracks. This would also make any attempt at additional environmental management extremely expensive in terms of both access and personnel costs.

The twenty-first century troopers are therefore in a very different situation from the 'one, two, three' featured in 'Waltzing Matilda'. In that situation, the troopers were using the force of the law to support production values that were concretised in the form of property, namely the jumbuck and the (admittedly leasehold) land belonging to the squatter on which the swagman was, arguably, trespassing. At Ningaloo, the troopers' task is to support the conservation values legally enshrined in Park designation and other environmental laws and regulations. This means, that the contemporary troopers are potentially opposed to the squatter insofar as the squatter's production values clash with conservation regulations. Furthermore, the consumption values of the 'swagmen' who seek a wilderness and, importantly, a largely unregulated, camping experience aligns them with the 'squatters' for whom the campers provide a valuable income (production?) stream. The waltz goes on, but some of the dancers may well have changed partners.

The story: tourism and heritage waltzes at Ningaloo

As indicated above, tourism development began in the late 1960s, when a hotel was constructed adjacent to the reef at Coral Bay. Here, as on the state's south coast during World War Two (Selwood et al. 1996), American servicemen were among the pioneer tourists, camping and fishing in the Ningaloo dunes. As the roads from Perth to the north and, indeed, around Australia were sealed in the 1970s and 1980s, the area became increasingly accessible. Furthermore, as more and more of the region's land

and sea areas were acquired for conservation purposes, its natural attractions became more widely known. Perhaps the most notable and exotic example of this was the publicisation of the annual whale shark migration to Ningaloo reef – and of the local potential for swimming with these animals – in Japan in the 1990s (Davis et al 1997).

In the terms of Butler's (1980) model of a tourism life cycle, this area therefore moved from 'exploration', through 'involvement' and into the 'development' phase within a few years and has, in little more than three decades reached a 'critical range of elements of capacity' if the region's attractions are to be preserved for future tourists and, indeed, for future generations.

Two issues in particular illustrate the extent to which these capacity limits have become critical, firstly the proposal to develop a major resort at Maud's Landing (the site of the early wool exporting jetty) near Coral Bay and secondly the plans for the management of camping along the Ningaloo coast.

Coral Bay/Maud's Landing

In the words of the Western Australian Planning Commission (2004, 107) 'Coral Bay has developed as a tourism settlement in a relatively ad hoc manner'. This site, which was excised from Cardabia pastoral station, had no town water supply and for this reason the local Shire opposed residential development there as long ago as 1973. Nevertheless, by 2004, it had grown to accommodate 25 tourism businesses, with at least 1848 beds and employing at least 150 staff (a level well beyond the bed capacity approved by the Shire of Carnarvon Planning Department). Its current capacity constraints are serious and immediate. The existing effluent disposal facilities of septic tanks, leach drains and evaporation ponds are totally inadequate and the State government is currently installing sewage treatment facilities. The service station and its fuel tanks located within the storm surge line. The staff accommodation is seriously substandard, being little more than a campsite. The airstrip is wrongly aligned with reference to the prevailing winds and fails to comply with Royal Flying Doctor Service requirements. Significant expansion at what is currently the only tourism node adjacent to the reef is therefore unfeasible without major infrastructural improvements.

In these circumstances, it was unsurprising that both Coral Bay's environmental problems and the negative impacts of unmanaged camping were cited as factors justifying the development of an alternative 'Coral Coast Resort'. The first proposals for such a development at Maud's Landing were made in 1987, the year in which the Marine Park was gazetted. However, the stock market crash, also in that year, intervened and the developers were unable to obtain financial backing for the 2,500 bed project until 1993. In 1995 the (Liberal-National) state government signed a heads of agreement with the developers to allow detailed planning of the project and, in 2000, work commenced between the developers, the Department of Land Administration and the Western Australian Tourism Commission on a Land Development Agreement, albeit for a resort of reduced capacity to that originally envisaged.

This proposal was clearly intended to cater for consumption values. At the same time, however, these proposals were generating considerable and high profile controversy from conservationists. A 'Save Ningaloo' campaign was launched in Perth, where it was associated with and received political and media advice from a highly successful pressure group campaigning against residential development at a metropolitan beach. Their cause attracted the support of national and international celebrities such as author Tim Winton, actor, Toni Colette and sports personalities Mick Malthouse and Luc Longley. At the 2001 State election, environmental issues in general experienced an unusually high profile, the opposition Labor Party gained a significant and unexpected victory and Green Party members gained a pivotal position in the upper house of parliament. In these political circumstances and in what was portrayed as a triumph for conservation values, the state government rejected the Maud's Landing proposal in 2003.

The management of coastal camping

The rejection of the Maud's Landing development re-turned the spotlight from the potential environmental impact of the resort back on to the actual impacts of the wilderness campers. In an email response to material on the saveningaloo.org website, world renowned marine biologist Karen Edyvane argued (in reference to the Maud's Landing development) 'Much better to have a single, integrated, environmental management system for a single development – than try to minimise the environmental impact of dozens of smaller developments over a large area.'

In an attempt to seek a compromise between the single resort and the 'dozens', if not hundreds, of separate campsites along the reef, the State government's response was to produce a major planning document 'Carnarvon-Ningaloo Coast: Planning for Sustainable Tourism and Land Use' (WAPC 2004). Under this scheme, the Western Australian Planning Commission will place the coastal area from Carnarvon to beyond Exmouth under a Regional Interim Development Order while a 'Region Scheme' is devised for it. This scheme will allow for more than a doubling of the tourist capacity at Coral Bay, with concomitant attention to such issues as waste disposal, water supply and staff accommodation and will seek to redistribute campers and recreationists using the coastal strip of pastoral and Defence Department land. The framework suggested for this in the strategy (WAPC 2004, 62) is for a 500 visitor and 200 visitor tourist nodes, an 'ecolodge node' for 100 visitors and eight camping nodes. It also suggests that the southernmost pastoral station, Warroora, may be suitable as a 'homestead tourism node' and acknowledges the possibility of some managed and monitored dispersed camping.

While it appeared that proposals of this kind could not be finally implemented until the revised pastoral leases, with the two kilometre coastal excisions were implemented in 2015 and that 'in the short term, much of the camping will continue to occur on the pastoral leases. Management of the camping will occur through a partnership with pastoral leaseholders, relevant State agencies, local government and visitors camping along the coast.' (WAPC 2004, 131) the dance has, in fact, continued at an accelerated tempo. Some pastoralists have already commenced negotiations with the State government which give them the entitlement to a 'development node' in return for the early surrender of their coastal excision.

Conclusion

The debates and conflicts over the development and/or preservation of the North West Cape-Ningaloo region clearly illustrate the differing views of the various protagonists on both the local environment and on the various heritages that it can be taken to represent. But all three groups experience some internal conflict. For the 'swagmen', an untrammelled camping, fishing and swimming holiday in a spectacular beach setting is both a perfect way to consume this remote environment and a cherished element of an idealised Australian lifestyle. Yet many of the campers surveyed by Remòte Research expressed their concerns over the environmentally insensitive behaviour of some of their fellow recreationists and their fears that, without greater formalisation of environmental management and conservation regimes, their holiday experience could become unsustainable. For the pastoralists, their choice of occupation could be said to have provided them with a uniquely independent and a nationally iconic means of making a living. But the properties that they lease have limited and frequently declining productive capacity – certainly in monetary terms given recent downward trends in wool prices. Involvement in tourism, either through 'station stays' or through wilderness camping can create additional income streams which can permit them to maintain many elements of their traditional lifestyle. For the 'troopers' their primary concern is the conservation of vulnerable ecosystems, species and landscapes. But a major reason for this conservation effort is to facilitate the consumption of these entities for educational and recreational purposes. For all the actors at Ningaloo, the roles that they play in the twenty-first century are therefore more complex than those of the 'Waltzing Matilda' originals.

Another significant difference between the story told here and Patterson's ballad concerns the networks in which these actors operate. In 'Waltzing Matilda' all the actors were local. Even the swagman was a rural dweller, allegedly a shearer travelling between properties, and possibly 'down on his luck'. In the recent disputes over Ningaloo, external and, frequently, urban influences and decisions predominate. Most of the campers are from Perth and the south of the state, as were the developers who sought to construct the Coral Coast Resort. The placing of a Regional Interim Development Order over the area has taken planning powers away from the Carnarvon and Exmouth local authorities and ceded them to the State government. While Ningaloo was an important environmental factor at the 2001 election, metropolitan concern over the fate of the state's southern forests was probably the key issue determining its final outcome.

Particularly after the pastoral lease revisions occur in 2015, it is likely that the natural heritage values of the area will be increasingly privileged over the cultural heritages/lifestyles of the pastoralists and the wilderness campers. Of the three reefside pastoral stations, one (Ningaloo) is proposed for near-complete resumption into the National Park and another (Warroora) is suggested as a 'Homestead Tourist Node'. The third (Cardabia) is also negotiating for a tourist node and may otherwise become unviable with the excision of the more productive two kilometre coastal strip. Indeed, the best hope for many pastoral stations may be to become part of the heritage industry (Hewison 1987) rather than to operate in their traditional manner even though this may require significant new skills acquisition and attitude shifts. The wilderness campers are also likely to find their holiday lifestyle circumscribed as conservation values increasingly dictate the ways in which they are able to consume the environment of this area.

While this might suggest that natural, and even scientific, heritage and conservation values are likely to win out locally over the cultural heritages and the production and consumption values of the pastoralists and campers respectively, this is something of an oversimplification. Tourism is a far more significant revenue generator and employer in this area than is agriculture. And most tourists visit this area because of its natural attractions. Conservation of these attractions therefore preserves an environment which tourists seek to consume and expenditure by these tourists fuels the area's (productive) economy.

The key factor which has transformed this region over recent decades has been a very geographical one, namely the massive reduction in its isolation. For wilderness campers, for outback pastoralists and for pristine ecosystems isolation is frequently a precondition of their unaided survival. On the evidence of its recent planning documents, in this case the government is choosing to aid the survival of the ecosystems: a choice which will inevitably impact upon the social sustainability and the heritages of the other two groups, but without, it is hoped, totally depriving them of their preferred lifestyles.

References

Anderson, K. (1999) 'Reflections on Redfern', in Stratford (ed.).

Armstrong, G. (2003) 'End of "Grass Castle" Kings', *The Sunday Times,* 3 August, 17.

Barthes, R. (1981) 'Semiology and the Urban', in Gottdeiner and Langopoulos (eds).

Blainey, G. (1966) *The Tyranny of Distance* (Melbourne: Sun Books).

Bolton, G. (1981) *Spoils and Spoilers: Australians Make their Environment 1788–1980* (North Sydney: George Allen and Unwin).

Burnley, I. (2005) 'Generations, Mobility and Community: Geographies of Three Generations of Greek and Italian Ancestry in Sydney', *Geographical Research* 43:4, 379–392.

Burnley, I. and Murphy, P. (2004) *Sea Change: Movement from Metropolitan to Arcadian Australia* (Sydney: UNSW Press).

Butler, R. (1980) 'The Concept of a Tourist Area Cycle of Evolution: Implications for Management of Resources', *Canadian Geographer* 24:1, 5–12.

Carlsen, J. and Wood, D. (2003) *Assessment of the Economic Value of Recreation and Tourism in Western Australia's National Parks, Marine Parks and Forests* (Gold Coast, Queensland: Sustainable Tourism CRC).

Cloke, P. and Little, J. (eds) (1997) *Contested Countryside Cultures: Otherness, Marginalisation and Rurality* (London and New York: Routledge).

Connell, J., King, R. and White, P. (eds.) (1995) *Writing across Worlds: Literature and Migration* (London and New York: Routledge).

Davis, D., Birtles, R., Valentine, P. and Cuthill, M. (1997) 'Whale Sharks in Ningaloo Marine Park: Managing Tourism in an Australian Marine Park', *Tourism Management* 18:5, 259–271.

Durack, M. (1959) *Kings in Grass Castles* (London: Constable).

Gottdeiner, M. and Langopoulos, A. (eds) (1981) *The City and the Sign: an Introduction to Urban Semiotics* (New York: Columbia University Press).

Gunn, A. (1908) *We of the Never-Never* (London: Hutchinson).

Heathcote, R. (1965) *Back of Bourke: A Study of Land Appraisal and Settlement in Semi-arid Australia* (Carlton: Melbourne University Press).

Herbert, X. (1938) *Capricornia* (Sydney: Angus and Robertson).

Hewison, R. (1987) T*he Heritage Industry: Britain in a Climate of Decline* (London: Methuen).

Holmes, J. (2002) 'Diversity and Change in Australia's Rangelands: a Post-productivist Transition with a Difference?', *Transactions of the Institute of British Geographers* 27:3, 362–384.

Holmes, J. (2006) 'Impulses towards a Multifunctional Transition in Rural Australia: Gaps in the Research Agenda', *Journal of Rural Studies* 22:2, 142–160.

Jones, R. (1995) 'Far Cities and Silver Countries: Migration to Australia in Fiction and on Film', in Connell, King and White (eds.).

Jones, R. (2001) 2001 'Social Sustainability in the Western Australian Urban Planning System: State Planning Strategies to 2029', in Singh (ed.).

Keys, L. (2006) 'Decision Time for Cattle Stations', *The West Australian* June 8, Liftout 12.

Remote Research (2002) *Summary Report on the Findings of Surveys of Unmanaged Camping in the North West Cape Region of Western Australia* (Perth: Remote Research).

Reynolds, H. (1990) *The Other Side of the Frontier: Aboriginal Resistance to the European Invasion of Australia* (Rev. ed.) (Ringwood, Vic: Penguin).

Selwood, J., Curry, G. and Jones, R. (1996) 'From the Turnaround to the Backlash: Tourism and Rural Change in the Shire of Denmark, Western Australia', *Urban Policy and Research* 14:3, 215–226.

Shaw, B. and Jones, R. (eds) (1997) *Contested Urban Heritage: Voices from the Periphery* (Aldershot: Ashgate).

Shute, N. (1950) *A Town Like Alice* (London: Heinemann).

Shute, N. (1956) *Beyond the Black Stump* (London: Heinemann).

Singh, R. (ed.) (2001) *Urban Sustainability in the Context of Global Change* (Enfield, New Hampshire, USA and Plymouth UK: Science Publishers).

Stratford , E. (ed.) (1999) *Australian Cultural Geographies* (Melbourne: Oxford University Press).

Tonts, M. and Grieve, S. (2002) 'Commodification and Creative Destruction in the Australian Rural Landscape: the Case of Bridgetown, Western Australia', *Australian Geographical Studies* 40:1, 58–70.

Tunbridge, J. and Ashworth, G. (1996) *Dissonant Heritage* (London: Wiley).

Western Australian Planning Commission (2004) *Carnarvon-Ningaloo Coast: Planning for Sustainable Tourism and Land Use* (Perth: Western Australian Planning Commission).

Chapter 7

Fixed Traditions and Locked-up Heritages: Misrepresenting Indigeneity

Wendy Shaw

Introduction

This chapter critiques the notions of 'tradition' and 'heritage', as they are applied to Indigenous peoples in Australia. I first examine the construction of 'traditional' Aboriginal gender relations within the non-Indigenous Australian (Westminster) legal system, where post/neo-colonial understandings about the Aboriginal 'other' have become enforced in law. I then consider the certainly popular and apparently general tendency for Indigenous heritages to be locked within archaeological pasts. I contend that, through these processes, constructions of 'tradition' and 'heritage' have contributed to an ongoing, but mostly unspoken, project of neo-colonialism.

Federal Indigenous Affairs Minister, Mal Brough, recently called for Indigenous self-determination to be scrapped, and for an end to be put to the discretionary use of traditional law in the trial and sentencing of Indigenous defendants.

> It is a fact that there is rampant Indigenous violence and I think the Australian society needs to face up to that reality, accept it, and then also accept as a nation that we aren't going to accept this any longer ... there has been far too much play by so-called [I]ndigenous leaders who hide behind the veil – the thin veil – of cultural sensitivity, when the reality is this is a total disaster for the people involved and the only cultural sensitivity here is that if we stand by and allow people to use these sort of excuses to allow their crimes to continue, then we have failed. (*Sydney Morning Herald*, 16 May 2006)

Mal Brough's call to 'scrap' Indigenous self-determination and legal reference to traditional/customary law occurred in response to the public release of a dossier on domestic violence and abuse in Aboriginal communities in May 2006. In this dossier, the Central Australia Crown Prosecutor, Nanette Rogers, brought the issue of Indigenous violence, and particularly sexual violence against children, to the attention of the wider Australian public. Brough's response was swift. But his reaction was not completely in step with what he had, perhaps, imagined would be the overwhelming reaction, namely that Australians, overall, would reel in horror at this news.

The sad reality is that the kinds of violence reported, and its reportage to the public in Australia and abroad, is far from new. What Brough did not account for was the sustained, and general throwing up of hands at the ongoing horrors experienced in Aboriginal communities. Other, 'ordinary' Australians have become almost immune to hearing about communities gripped by poverty and its associated vices. And, far from acknowledging and *responding* to the ravages of dispossession and the consequent attempts to cling to vestiges of cultural practice, it is far easier to blame Aboriginal people themselves for the kinds of violent events portrayed in the media in May that year. A friend, an Aboriginal psychologist, mentioned to me at the time that she had become all too familiar with a preoccupation (by non-Aboriginal counsellors), who simply had to ask her: was child sexual assault a *traditional* (my emphasis) practice in 'her culture'? Her responses have varied but, in her frustration, she has resigned herself to a simple quip, along the lines of 'oh no, you lot (colonisers) brought that one with you'.

This chapter considers some of the wider conceptualisations of 'tradition' and 'heritage', which tend to remain discursively fixed when they are applied to settler understandings of Indigeneity. In this chapter, I will specifically refer to peoples who exist at the coalface of the idea of 'tradition' and yet, paradoxically, remain largely external or marginal to the general orbit of 'heritage', its meaning and (re)productions, and in particular, its industries. Settler concepts of Aboriginal peoples are, all too often, constructed around understandings which conflate tradition and authenticity and thus perceive any changes in Aboriginal behaviour patterns or location as being inauthentic in cultural terms. As I have documented elsewhere, the Aboriginal settlement known as *The Block*, in the inner Sydney suburb of Redfern, is constructed as 'out of place' because it represents a (spatial) break with tradition. It is an urbanised form of Indigeneity, rather than being authentically – at least in popular terms –'outback', and it is therefore unpredictable, dangerous and a wild 'failed human experiment' that cannot belong in that location (Shaw 2000). *The Block* is widely regarded by non-Indigenous Australians[1] to be troubled, not because of the conditions of extreme poverty and dispossession experienced by its inhabitants, but because in the city, so the narrative goes, Indigenous Australians simply cannot cope. Rather than being 'at one' with their culture(s) when they are in a metropolitan setting, civilisation sends them 'crazy'. Such racialised beliefs about Aboriginal Australians can then be used to rationalise the incidences of violence referred to above.

Somewhat ironically, the converse is true with regard to popular understandings about Aboriginal 'heritage', which tends to remain locked into a slightly different settler understanding of 'tradition'. In this version, Aboriginal traditionalism is somewhat romantically viewed from the present, and it is from the depths of history that Indigenous traditions are seen to emerge. When Aboriginal heritage does manifest itself in the present, its remnants – in the form of cultural artefacts – are embraced for their archaelogical value, which is seen as conveying their authenticity as 'Aboriginal'. So, in the Australian context, Aboriginal traditions and heritages are widely viewed as pre-colonial; as objects and practices from before the colonial record. They are therefore rendered pre-historic. This image of Indigenous tradition (and its heritages) means that it is seen as remaining intact only in the very distant

---

1   It must be noted that on occasion, there have been comments made by high-profile Aboriginal spokespeople against *The Block* as well. Of course, the media have seized on such comments as 'truth'.

past and as being destroyed both by and in the present. It is therefore perceived as being incapable of evolving and as being associated with cultural practices that are anachronistic (and unacceptable) in the present.

Aboriginal 'heritage' is tied to these sorts of understandings about tradition because it too has no place in the present or in recent history *except* for the purposes of providing historical evidence, or for consumption of the exotic (Thomas 1994).

Drawing on a range of research, from legal, urban and Indigenous geographies, my discussion here considers these concepts of Aboriginal 'tradition' and 'heritage' in urban and non-urban Australia. Armed with information gathered from a range of research endeavours, and using a full range of methodologies or multiple methods approaches available to the pursuit of human geography today, my task is to demonstrate how these concepts of 'tradition' and 'heritage', are being applied – and *not* applied – to Indigenous peoples for the purpose of furthering a (neo)colonial project of othering Indigeneity. I begin by tracing the use of what Attwood and Arnold (1992) once dubbed the 'Aboriginalisms' that are used within the Australian legal system to construct 'traditions'. I then turn to the ever-evolving constructions of 'heritage' currently being applied to the swiftly gentrifying urban environment of inner Sydney. In this context, 'heritage' is celebrated but it speaks of a very specific, and exclusive history of British settlement in which the architectural artefacts of colonialism have become the desirable remnants, the *post*colonial heritage, of colonial pasts. All other heritages, of Indigenous or migrant settlement, do not fit within such understandings. In fact, they are contrary to them, and are thus widely regarded as *anti*-heritage. I conclude this chapter with a discussion of the contentiousness of the meanings of 'heritage' and 'tradition', particularly as they are currently ascribed to and prescribed for postcolonial Indigeneity in Australia (and beyond).

Tradition(alism)

The concerns raised by Nanette Rogers, and the response by the Federal Indigenous Affairs Minister, Mal Brough (above), link back to a decision made in the late 1960s, to include legal reference to Aboriginal Customary Laws in cases of violence committed against Aboriginal women (and children), by Aboriginal men. The consolidation of Aboriginal politics at that time, including the belated granting of Australian citizenship to the Aboriginal population, had highlighted the inordinately high incarceration rates of Aboriginal people in Australia. An initiative, introduced to legal process in an attempt to alleviate these escalating incarceration rates,[2] was the incorporation of Aboriginal Customary Laws (or 'tribal laws') into the Australian legal system when cases involving Aboriginal defendants were being heard. This included an acknowledgment of traditional cultural practices that could, at times, include physical punishment of the offender (Crawford and Hennessy 1982). More broadly, the recognition of Aboriginal Customary Laws, and their inclusion into the (largely

---

2   Regardless of the Royal Commission to investigate Aboriginal deaths in custody (and incarceration rates) in 1987, by 2000, the rate of imprisonment for Aboriginal and Torres Strait Islander people had reached approximately 14 times that of the non-Indigenous population (ABS 1994, 2002).

non-Aboriginal) Australian legal system, signalled a step towards Aboriginal self-determination through the acknowledgment, and the inclusion, of certain Indigenous cultural practices in the legal mainstream. Ultimately, however, the process proved to be flawed because it perpetuated and even publicised misunderstandings about Aboriginal traditional practices.

During colonisation/dispossession, acts of violence, bannered as 'Aboriginal tradition', were thoroughly outlawed with the gradual and expansionary imposition of the British Westminster System. Ritual punishments for breaking Aboriginal Customary Laws still do occur occasionally, usually in remote parts of Australia. However, by the time of the Laverton Royal Commission[3] (1975–76), relevant criminal statutes were amended to include some 'tribal' (Aboriginal Customary) Laws. However, as Marguerite and McNamara (2000, 16) observed, 'the practice of taking "payback" into account [in law] continues to operate on the somewhat unpredictable basis of common law sentencing discretion, without a solid legislative foundation'. It is therefore far from standard practice.

This incorporation of Aboriginal Customary Laws into legal process was highly contentious in cases involving the 'tribal punishment' of Aboriginal women and children. The realities of domestic violence are very difficult to reconcile, particularly when they have the potential to be touted as being some kind of traditional practice. According to Marguerite and McNamara (2000), 75.4 per cent of Indigenous women victims of murder were killed by their intimate partners (compared with 54 per cent for non-Indigenous women victims), and Indigenous women were victims of crime at an annual rate of 11.7 per 100,000 population (compared with 1.1 per 100,000 population for non-Indigenous women). Marguerite and McNamara (2000, 8) also reported that:

> Incidents of rape and violence [against women] are occurring in association with other situational factors such as alcoholism, spousal violence, poor community facilities, lack of education, support services and resources. (Commonwealth Attorney-General's Department, 2000)

Additionally, legal systems are not neutral forums, and the Australian system has been widely criticised for its displays of gender biases.[4] In this context, it is difficult to forget comments made by high-profile legal professionals such as Judge Bollen, who claimed that 'a husband could use rougher than usual handling to force a reluctant wife to agree to sex' (*Sydney Morning Herald*, 21 April 1993) and Judge Bland, who believed that 'No, often subsequently means yes' referring to women saying 'No' to unwanted sex (*Sydney Morning Herald*, 6 August 1994). Similarly, Judge O'Bryan's infamous gaffe that 'a woman would be less traumatised by rape, and having her

---

3   The Laverton Royal Commission in Western Australia investigated clashes between police and Aboriginal people at Laverton and Skull Creek in December 1974 and January 1975.

4   Which stem, in part, from historical bias in law. For example, it was not until 1989 that the possibility of 'rape' (sexual assault) in marriage was accepted in law in Queensland (Bulbeck, 1993). See also the collection of papers in the first edition of *Australian Feminist Law Journal*, 1993.

throat cut because she was unconscious at the time [than a fully conscious victim]'
(*Sydney Morning Herald*, 6 August 1994), and Michael Rozens QC's conclusion that
'the rape of a prostitute was a lesser crime than that of a nun' (*Sydney Morning Herald*,
22 April 1994), were both telling. Although these examples are not representative of
all legal practice, they certainly pointed to a troublesome discursive environment,
which seemed out of step with contemporary understandings of such matters. Add a
layer of old-fashioned racialisation to this already unseemly and dated gender bias,
and there is little wonder that Indigenous women tend not to trust the legal system
(Atkinson 1990).

With the existing over-representation of Aboriginal people within the criminal
law system, and the added burden of the impact of Mandatory Sentencing Laws,[5]
Aboriginal women have not only suffered the burdens of gender bias within the legal
system, but they have done so at the hands of 'white law'. As I have documented
elsewhere (Shaw 2003), there has been a history of silencing Aboriginal women in
legal proceedings. Their versions of events (that is, their evidence) and their views
on the constitution of traditional practices, have rarely been considered. The wider
implications of the courtroom experiences summarised below point to more general
neo-colonial processes which include the high level of Aboriginal incarceration rates,
and a persistent belief in the violent 'nature', or genetic 'race trait', of Indigenous
Australians.

Postcolonial legalities of 'tradition'

The central dilemma in debates about the inclusion of 'tribal' or customary laws
continues to relate to the powerful immovability of notions of Indigenous 'tradition'
on the one hand the counter logics of basic and universal human rights on the other.
There is some acknowledgement (which does not stretch to the realms of political
commentary) that, although violence against women and children should never be
excused, in the case of Indigenous Australians, complexities arise from the ongoing
(post)colonial oppression and racism experienced by Aboriginal people (Larbalestier
1990; Goodall and Huggins 1992; Atkinson 1996). The dominant system of law in
Australia has thus played a forceful role in the ongoing production of (post)colonial
oppressions (Cuneen 1990).

Throughout my analysis of a series of court cases (Shaw 2003) where Aboriginal
Customary Laws have been invoked in defence of violence committed by Aboriginal
men against Aboriginal women or children, stereotypical constructions of 'race',
cultural authenticity and gender relations, persisted. I used a simple technique of
following the 'footprints' of the entry point/production, transmission and reception
of knowledge (Mani 1990) about Aboriginal Customary Laws, within legal
discourses. The discourses analysed included non-Aboriginal testimonies and, at

---

5   Mandatory Sentencing Laws mean that any person 17 years and older, found guilty of
property offences is liable to serve a minimum of 14 days in prison for a first offence (March
1997 in the Northern Territory, though Western Australia introduced similar laws in November
1996). A likely consequence of Mandatory Sentencing is an increase in the rate of Aboriginal
deaths in prison. (http://www.aph.gov.au/library/intguide/law/mandatorychronology.htm).

times, testimonies extracted from defendants, and members of their communities. The select sources of knowledge about Aboriginal traditions came from 'expert witnesses', and included select anthropological data, discussions with tribal Elders and the opinions of the accused. Where no other 'expert witness' was available, legal practitioners occasionally provided this cultural information. Using this technique, I unveiled the inadequacies, misdirections and/or contradictions that occurred in the legal translations of *tradition*.

A precedent-setting case in 1976 illustrated how legal narratives were used to construct the notion of 'tradition'. In this case 'tribal punishment' (under Aboriginal Customary Law) was included as part of the sentence because an initiated Pitjantjatjara[6] man had killed a woman (who had insulted him by mentioning, while intoxicated, tribal secrets that women are not supposed to know). He hit her with a stick and a bottle and she died because of the injuries. In this case, the judge accepted that there was sufficient evidence of provocation by the woman for tribal punishment to be imposed and that it was a 'traditional killing', a decision based on the assumption that the woman (who was 'not respected') was guilty of breaching traditional law (i.e. she was found guilty of provocation) and that she was therefore punished. The judge also accepted that, because the woman's punishment had been overly severe, the defendant would also face 'tribal punishment' and that this needed to be taken into account in his sentencing. Controversially, no sentencing or imprisonment resulted. However, the tribal Elders, also did not punish the defendant because, as far as they were concerned, the crime had not occurred under any 'tribal' circumstances. They had not been consulted over the original sentencing and were deeply perplexed by the court's decision. This troubled and highly fraught case did, however, establish a legal precedent for the use of Aboriginal Customary Laws in sentencing (Ward 1976). What it did not establish was the existence, or any reference to the existence, of Aboriginal Customary Law or custom about the violent punishment of women, nor that violence against women was acceptable practice in Aboriginal Australia. Yet such 'understandings' are now popularly established.

In other words, the evidence (or indeed *lack* of legal evidence) suggested that the Aboriginal Customary Laws used in these court cases were largely fictitious (cf. Mahoney *et al.* 1993), and were created through legal (mis)understandings about Indigenous culture(s) and traditional practices. Such fictions have found their way into the establishment of *stare decisis* (precedent). And it is the reproduction of such fictions that leads to the kinds of sporadic outcries that were cited at the beginning of this chapter.

The creation of such fictions as 'culture' or 'tradition' is not solely the domain of legal process. Following Said's notion of 'Orientalism' (1978), Attwood and Arnold (1992, iii) have invoked the term 'Aboriginalism':

Aboriginalisms produce the reality … imagined by influencing government policies and practice which have, in turn, determined Aborigines' terms of existence – racialising the Aboriginal social body and so making Aborigines of the Indigenous population.

6　Pitjantjatjara people are from a remote part of Central Australia. Pitjantjatjara is also an Indigenous language group.

Aboriginalisms have certainly been active in a range of contexts. They are invoked in the performances of stereotyping from the level of federal governance in this country (as exhibited by Mal Brough), to the level of the street, every day.

Through its less than satisfactory attempts at exhibiting cultural sensitivity through the process of incorporating what were believed to be 'traditional practices', the Australian legal system has simply reinforced and consolidated its (neo)colonial power base. These geographically wide-ranging and temporally sporadic, yet fixed and universalised, understandings of Aboriginality have resulted in its ongoing construction as barbaric and in need of protection from *itself* while, at the same time, it is rendered, in the feminine, as wild, promiscuous and provocative.

In the meantime, the inclusion of Aboriginal cultures, of *actual* Customary Laws and practices into legal process, remains controversial and such attempts are occasional and, as a result of the recent controversies, largely endangered. The seemingly immovable contest has been fixed, through legal discourses, on the contest between perceived (abhorrent) cultural practices versus human rights. At the same time, progress towards Aboriginal self-determination continues to be hindered because the reputation of a set of different laws has been almost irretrievably tarnished. Such laws, and their capacity to help alleviate the incarceration rates for Aboriginal peoples, increasingly sit outside of the domain of the Westminster System of Law in (post)colonial Australia.

In the case of a concept inextricably tied to 'tradition', quite a different story emerges with designations of 'heritage'. The kinds of understandings evinced in many listings found in heritage registers[7] depict the cultures of Indigenous peoples as non-evolving, and as being fixed in prehistory. The following section draws on contemporary constructions of heritage in inner Sydney. It provides examples of the use of 'heritage' designations as part of a wider project of exclusion of Aboriginal peoples. Their histories, particularly since colonisation, are simply not represented in these contemporary imaginings of heritage.

Heritage (impulses and constructions)

In this next section, I consider some of the motivations behind the use of the concept of heritage, its selective designation, its expanding embrace, and its protection (Shaw 2005). Examples are drawn from one of gentrification's 'final frontiers' (cf Smith 1996), where a specifically designated Aboriginal place known as *The Block*, 'Aboriginal Redfern' or 'Eveleigh Street' sits in increasingly uneasily within the transforming city that surrounds it (Figure 7.1).

As Sydney takes its place as a globalising metropolis, a physical and economic renaissance is occurring. The Central Business District (CBD) is awash with apartment development, and the old housing areas nearby, which were shunned during an earlier

---

7   See, for example, the Australian Heritage Council (www.ahc.gov.au/), and also the National Register of Historic Places, Hawai'I County (www.nationalregisterofhistoricplaces. com/HI/Hawaii/state2.html) where 'heritage' is largely conceived of as colonial buildings and armaments. 'Native' Hawaiians (Kanaka Maoli) are represented by a few archaeological sites.

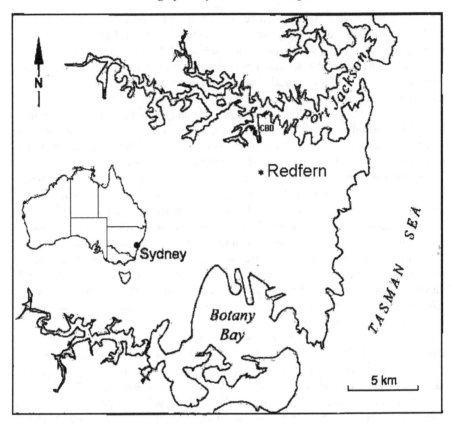

Figure 7.1   Map of Sydney

era of suburbanisation, are now increasingly valorised as 'heritage' housing areas. The graceful terrace houses of the inner city, with their nineteenth century Victorian architecture, have become highly desirable. The former industrial areas of Pyrmont, Ultimo and Chippendale, are also changing, with many old warehouses and factories now converted into apartments. However, this 'heritage' is far from inclusive and, even with its formalised heritage identity,[8] the Aboriginal settlement of *The Block* cannot participate in the glory of the locality's heritage designation.

The neighbourhoods that surround *The Block* have been transformed swiftly and dramatically from undesirable to desirable urban spaces, and their 'heritage' associations are central to their newfound status (Figure 7.2). The Redfern area is also a place with a history of struggle. In the 1960s, the neighbourhood surrounding *The Block* was threatened with complete subsumption by The University of Sydney. Then, in the 1970s, in another struggle over territory, a patch of Darlington became

---

8   On 25 October 2000, the Australian Heritage Commission announced that *The Block* had been listed on the National Heritage Register of Australia as a site of Indigenous significance.

The Block.[9] Colonisation of the Sydney region had dispossessed the Gadigal/Eora people of their country, as they were quickly and thoroughly exiled during the early days of British settlement (Reynolds 1996). Then, from the 1930s, when the earliest migrations (back) to what had become the city of Sydney began, urban settlement has also worked to displace Aboriginal people. The recent onset of gentrification has meant a new set of responses to the existence of *The Block*. It is now an extremely valuable site.

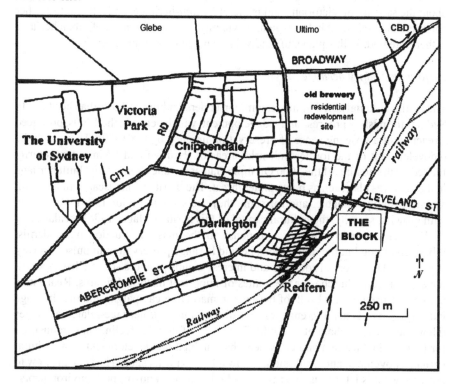

Figure 7.2   Map of Redfern area

Meanwhile, the predominant script of *The Block,* in popular imaginings, is that it is spiralling into self-inflicted decline. The mostly non-Aboriginal gentrifiers are drawn to this previously avoided part of the city to purchase heritage housing that is still, comparatively, reasonably priced. As the heritage imaginary matured, it has become part of the cultural capital of gentrification. In 1996, Darlington was listed as a 'heritage and conservation' area (South Sydney Local Environment Plan, 1996) and

---

9   In 1973, during a storm of controversy, the then Prime Minister of Australia, Gough Whitlam, granted money for the purchase of a site of 70 Victorian terrace houses for inner-city Aboriginal people. The Aboriginal Housing Committee was formed to receive the grant to purchase and administer what became 'The Block'. This was a grand gesture by a new radical Labor government that had been elected during an era of massive social upheaval (Anderson 1993)

it is also listed on the Australian Heritage Commission's National Register (Database Number 001785, File Number: 1/12/033/0011). In sharp contrast, the Victorian terrace houses on *The Block* are run-down, derelict or even razed to the ground. Newly arrived gentrifiers find that *The Block* does persist as a place of Aboriginal settlement, regardless of popular beliefs and portrayals of its decline. Its high-profile street life and, at times, overt poverty, is witnessed by all who traverse the area by car, by train, or on foot from the railway station. *The Block* flouts an emergent convention about the sanctity of Victoriana and is therefore *anti*-heritage, and a threat to further commercial investment in heritage in its vicinity. The performance of urban heritage is therefore tied, in this place, to land values and gentrification motives.

Performing Sydney heritage

The British colonial past dominates common understandings of Australian heritage. In Sydney, the push of progress and redevelopment, which began in the 1960s, produced a reaction of a growing sense of preservation urgency by the 1970s. Then, with the redevelopment of the former industrial areas of Pyrmont and Ultimo, temporarily fuelled by the hosting of the 2000 Olympic Games, Sydney's industrial built heritage in the area was also seen as being under threat. The 'heritage' landscapes that remain are increasingly valuable, and for some they are sacred (Taylor 1994). Of course, with growing scarcity, the cultural capital of this heritage (Jager 1986) escalates.

For many, conservation is a response to the mass destruction of modernist urban rebuilding but the motivations behind heritage preservation impulses do vary (Crang 1994). Historical societies and interest groups promote specific heritages of, for example, old churches or the homes of important historical figures. Real estate marketeers use the 'hard sell' of heritage in marketing campaigns and the 'heritage industry', more generally, engages in a diverse range of heritage-related activities (Hewison 1987, Graham et al 2000). The depth of popular feeling and enthusiasm for heritage in Australia and elsewhere, has enabled the development of a 'cultural heritage movement' and professionals now advise on what should be kept and how to preserve what is left of just over two hundred years of (known) built environments.

Ken Taylor (1994), has identified a number of factors that have influenced the emergence of this movement, in Australia, which began with a surge in interest during the massive urban redevelopments of the 1960s and 1970s. A second influence was the resurgence of nationalism, in the 1970s. Thirdly, the development of heritage management as a profession and public recognition of its potential were boosted in 1988 when governments opened their coffers in support of the Australian Bicentennial. The Australian Heritage Commission (AHC), the AHC Act of 1975 and the Register of the National Estate and its associated legislation have provided legitimacy for the protection of (built) heritage. The reification of, for example, heritage tourism (Ashworth and Tunbridge 1990, Waitt and McGuirk 1996), and the recent rise of heritage awareness and protection through the gentrification movement, all point to a thriving local industry.

As a former British colony, Australia, notwithstanding its Indigenous (pre)history, is commonly regarded as lacking a recognisable, independent lengthy history upon which to build a national identity (Taylor 1994). 'We' are therefore creating our own (heritage/identity), based on elements of connection that were imported during the waves of 'migrations' that began a little more than two hundred years ago.

Whose pasts are 'heritage'?

Because heritage is more than a simple reuse of the past (Graham et al 2000), heritage impulses have frequently raised suspicions. Lowenthal (1985), Jager (1986) and Rosaldo (1989) have examined yearnings for the past as a form of power. Jager (1986) detailed the operation of class in heritage appreciation. As property developers know all too well, urban conservation can reuse history for the saleable purpose of manufacturing social distinction. It can also reflect a yearning for 'other' times and/ or cultures. According to Rosaldo (1989) 'imperialist nostalgia' is disguised as an 'innocent yearning' for pasts that were often brutal in terms of their domination of others. These complexities of race relations and colonial nostalgias (Jacobs 1992) are exemplified in Australia's 'innocent' yearnings for specific heritages. These heritages glorify colonial buildings, and bury Indigenous heritage(s) in prehistory. This dynamic of ancient Indigeneity, and temporally fluid non-Indigenous heritage, has been played out at the 'birthplace' of colonial Australia, called 'The Rocks'.

Since its 'restoration', The Rocks now represents an idealised and sanitised history of colonisation (Bennett 1993). The past has been (re)fabricated and cleansed of the marks that bear testimony to other, contradictory, aspects of the history of colonisation. 'The glittering façade ... functions as an institutional mode of forgetting' (Bennett 1993, 225). The new allegory is the ascent of 'a free, democratic, multicultural citizenry' (Bennett 1993, 227) with Aboriginality referenced only in 'traditional artefacts', which can be purchased from a craft shop. For Bennett (1993, 228), the notable lack of an Aboriginal presence reinforces the commonly held belief, and hope, that 'European civilisation' has tamed 'the natural' in this location. Indigenous occupation was removed as part of 'the natural' of the past (Jacobs 1996, Anderson 2000). This natural past has been overwritten by (non-Aboriginal) 'humanity', and the controversy of invasion lies silent beneath the re-written layers.

In inner Sydney's Redfern, by contrast, the Aboriginal presence is undeniably still present. *The Block* community consists of many kinship groups from around the country,[10] so there is no specific ancestral claim to this place. Its Indigeneity therefore, cannot be relegated to a pre-colonial past. Aboriginal Australia won this slice of the city of Sydney as part of a different history – a history of struggle and survival through an emergent 'black politics'. However, and regardless of *The Block's* heritage listing, the understandings of heritage that surround it remain fixed on old buildings, and their preservation.

Sydney's migrant experience has been similarly relegated to the domain of *anti*-heritage (Armstrong 1994, Lozanovksa 1994). Post-war migrants 'rescued the reputation of the terrace house as a place to live as well as restor(ing) its fabric' (Howe 1994, 155). First stage gentrifiers did the same things, but these migrant others are now blamed for defacing 'authentic' heritage. Real Estate agents associate

---

10 *The Block* is a strategic meeting-place for the most dispossessed, many of whom are from the *stolen generations* of Aboriginal peoples.

'tasteless' renovation such as the replacement of decorative (often rotting) timber windows with ugly aluminium, and (again, often rotting) timber floors with concrete (Research Interview, 4 March 1998). Heritage, on the other hand, is associated with restoration/replacement of timber windows and floors. The modifications that migrants from Southern Europe made to aging housing stock, to make them more liveable and culturally appropriate, now constitute a deficit in heritage capital. The layer of history that they added is now unwanted, and is usually removed. The preservation of intact Victorian architectural features (or, failing that, their replication), without the hindrance of other layers of history/defacement, have now become a priority for those who can afford them.

Meanwhile, around inner Sydney, the 'consumption circuit' (Jager 1986, 87) of heritage continues to expand. From grand Victoriana, to the inclusion of small Victoriana and old factories and warehouses, 'retro chic' and 'new-build' faux-heritage, have also gained status. Although not heritage per se (particularly in the case of faux-heritage) these forms of 'neo-archaism' (Jager 1986, 88) are expressed through new developments and building conversions. References can be found in 'heritage' brickwork, cobbled laneways, stone gutters and faux Victorian housing and these are preferred over non-heritage building designs in this part of inner Sydney.

As the meaning of 'heritage' continues to expand, and to become more rubbery and diverse, desires for built heritage, and its preservation are mainstreamed. Such desires prioritise the symbols of select pasts, of terrace houses and old industrial façades, over contemporary expressions of human diversity. In inner Sydney, resident activism mobilises to fight developments that are regarded to be 'tasteless' (non-heritage referenced) and/or dedicated to the housing of others, such as students from abroad. Such protectionist activities have assisted in the disengagement of the local heritage movement from, and concerns about, human diversity, yet it is this 'diversity' that is part of the lexicon of expanded heritage designations. Threats to old architectural diversity are monitored and protected by 'the community'. At the same time, desires to protect human diversity (such as the memory of those who may have toiled behind the protected façade, the migrant others, or the Aboriginal community around the corner) are sidelined by concerns about who will occupy new developments and the consequent impact of this on land and building values. The intertwining of heritage and taste percolate through the discourses of protest against such developments and these structures now represent the benchmark for *anti*-heritage design.

Darryl Crilley (1993) has identified how property developers promote diversity to enhance the mainstream appeal of developments rather than to appeal to a diverse market. Homogeneity can be concealed within the appearance of diversity, as expressed through, for example, heritage architecture. 'Diversity', in this case, is simply another consumable attribute for affluent tastes (cf Hage 1998, Bourdieu 1984) and rather than appealing to a range of people, it only does so to those with the necessary attributes (such as cash, class and/or ethnicity) to have membership in such a niche market.

This preoccupation with protecting symbols from the (neo-)colonial past(s), or allowing only those developments that are deemed tasteful to middle-class sensibilities, has enacted an architecture of denial whereby human diversity is

Figure 7.3   Bulldozing 'heritage'

denied, as the expanding orbit of heritage designations continues on its trajectory of exclusion. The deeply embedded desires to preserve (colonial remnants) and protect (white space) have become an escape from everyday realities of a colonial aftermath that has produced, for example, overt Aboriginal poverty and dispossession just around the corner from a tasteful 'heritage' built environments. The notion of heritage, as it is popularly conceived and as governments have legislated for it,[11] exhibits a certain consistency. Heritage remains commonly associated with old buildings and objects.

The unspoken heritage story around *The Block* is the history of colonialism, of encounters between Aboriginal and non-Aboriginal people. The heritage of *The Block*, its struggle for Aboriginal civil rights, is excluded from local heritage understandings regardless of its formal heritage status (Figure 7.3). There is no commemorative plaque, or acknowledgement that this site has finally been recognised as a site of cultural heritage. Heritage, as it is commonly conceived here, is part of an unspoken definition of 'community', of belonging. For the area around *The Block*, heritage remains architectural/artefactual. Where people are considered, it is the elegant lives of those who could afford High Victoriana or, in a more recent working of heritage, it is a partial legacy of (the housing of) the working classes, that is collectively

---

11 A 'Heritage and Conservation' Local Environment Plan for South Sydney was introduced in 1996, and the Heritage Council of NSW announced that a 'New State Heritage Register' for 'state icons' had been established (1998) by amendment to the 'Heritage Act' of 1977. At a national level the term 'National Estate' was adopted in 1972. There is the Heritage Commission, and the independent National Trust of Australia, established in 1950, which holds a classification of heritage register.

remembered (Boyer 1998). A yearning for 'more of the same' (pasts) is sometimes muffled in such imaginings (Bennett 1993, 235).

While Aboriginal people may be acknowledged as having heritage, such understandings are tied to the non-urban, cultural and ethnic pre-colonial homogeneity. The politics of the 'Black Capital' (*The Block*), of the unification of disparately dispossessed Aboriginal peoples,[12] is continually being written out of the evolving urban heritage imaginary. The non-Aboriginal, non-migrant gentrifier imagining(s) of heritage has excused itself from engagement with the urban histories of others.

Conclusions

In this chapter I have considered the notions of 'tradition and 'heritage', particularly as they are applied to Indigenous peoples in Australia. I have unpacked some of the construction of 'tradition' as expressed through settler understandings of Aboriginal gender relations, within the non-Indigenous Australian 'Westminster' legal system. In this context, post/neo-colonial understandings about the Aboriginal 'other' that were enforced in law persist and reinforce stereotypical beliefs about Indigeneity in Australia. In the case of 'tradition', beliefs about the brutality of 'race' and/or 'culture' are summoned from an imagined past, and have been fixed to Indigeneity, in the present.

Indigenous 'heritage', somewhat differently, remains discursively locked in archaeological pasts. In a process of reinvention, a time-line of what constitutes 'heritage' in inner Sydney starts with Victorian terrace houses. It has been expanded but not to the inclusion of Migrants, and their efforts to preserve and utilise crumbling housing stocks, the labours of the 'working classes', and, still less so, to the emergent 'black politics' of the 1970s, through the formation of *The Block*. Current preoccupations with (evolving) heritage designations have proved to be a useful way to deny the realities of class relations and neo-colonialisms. Because Aboriginal people and their associated places have been disengaged from mainstream experiences of more recent pasts, they have become museum-like objects (Wasserman 1984, 1994, Thomas 1994), or been disregarded altogether.

Conceptualisations of 'tradition' and 'heritage' thus tend to remain discursively fixed when applied to understandings of Indigeneity. Indigenous peoples are thought of as anchored to 'traditional' practices that are abhorrent or out-of-place in the contemporary world. At the same time, Indigenous heritage, as archaelogical and exotic, is glorified and revered. Postcolonial Indigenous heritage remains largely out of place within the expanding understandings of heritage, particularly in urban contexts. Aboriginal peoples remain, all too often, authenticated by fantasies from the not so distant colonial and far more distant pre-colonial pasts.

---

12 The now well recognised Aboriginal flag was conceived in 1971 as part of a politics of unification.

References

Anderson, K. (2000) 'The Beast Within': Race, Humanity, and Animality, *Environment and Planning D: Society and Space* 18, 301–320.

Anderson, K. and Gale, F. (eds) (1992) *Inventing Places: Studies In Cultural Geography* (Melbourne: Longman Cheshire).

Armstrong, H. (1994) Cultural Continuity in Multicultural Sub/urban Places, in Gibson and Watson (eds).

Ashworth, G. and Tunbridge, J. (1990) *The Tourist Historic City* (London: Belhaven Press).

Atkinson, J. (1990) 'Violence in Aboriginal Australia: Colonisation and Its Impact on Gender', *Refractory Girl*, 36, 21.

Atkinson, J. (1996) 'A Nation is not Conquered', *Domestic Violence and Incest Resource Centre Newsletter*, 3, August, 26–32.

Atkinson, R. and Bridge, G. (eds) (2005) *The New Urban Colonialism: Gentrification in a Global Context* (London: Routledge).

Attwood, B. and Arnold, J. (eds) (1992) *Power, Knowledge and Aborigines* (Bundoora, La Trobe University Press).

Bennett, T. (1993) 'History on the Rocks', in Frow and Morris (eds).

Bourdieu, P. (1984) *Distinction: A Social Critique of the Judgement of Taste*, (London: Routlege).

Boyer, M.C. (1998) *The City of Collective Memory: Its Historical Imagery and Architectural Entertainments*, (Massachusetts: MIT Press).

Crawford, J.R. and Hennessy, P. K. (1982) *Reference on Aboriginal Customary Law Research Paper No. 6A, Appendix: Cases on Traditional Punishment and Sentencing* (Sydney: Australian Law Reform Commission).

Crang, M. (1994) 'On the Heritage Trail: Maps of and Journeys to Olde Englande', *Environment and Planning D: Society and Space*, 12, 341–355.

Crilley, D. (1993) 'Architecture as Advertising: Constructing the Image of Redevelopment', in Kearns and Philo (eds).

Cunneen, C. (1990) 'Aboriginal-Police Relations in Redfern: With special reference to the 'Police Raid' of 8 February 1990', (Report Commissioned by the National Inquiry into Racist Violence, Human Rights and Equal Opportunity Commission, May 1990).

Frow, J. and Morris, M. (eds) *Australian Cultural Studies: A Reader* (Sydney: Allen and Unwin).

Gibson, K. and Watson, S. (eds) (1994) *Metropolis Now* (Australia: Pluto Press).

Goodall, H. and Huggins, J. (1992) 'Aboriginal Women are Everywhere: Contemporary Struggles', in Saunders and Evans (eds).

Graham, B., Ashworth, G.J. and Tunbridge, J.E. (2000) *A Geography of Heritage: Power, Culture and Economy* (London: Arnold).

Hage, G. (1998) *White Nation: Fantasies of White Supremacy in a Multicultural Society* (Sydney: Pluto Press).

Hedon, D., Hooton, J. and Horne, D. (eds) (1994) *The Abundant Culture: Meaning and Significance in Everyday Australia* (Australia: Allen and Unwin).

Hewison, R. (1987) *The Heritage Industry: Britain in a Climate of Decline* (London: Methuen).

Howe, R. (1994) 'Inner Suburbs: from Slums to Gentrification', in Johnson (ed).

Jacobs, J. M. (1992) 'Cultures of the Past and Urban Transformation: the Spitalfields Market Redevelopment in East London', in Anderson and Gale (eds).

Jacobs, J. M. (1996) *Edge of Empire: Postcolonialism and the City* (London: Routledge).

Jager, M. (1986) 'Class Definition and the Aesthetics of Gentrification: Victoriana in Melbourne', in Smith and Williams (eds).

Johnson, L. (ed) (1994) *Suburban Dreaming*, (Victoria: Deakin University Press).

Kearns, G. and Philo C. (eds) (1993), *Selling Places: The City as Cultural Capital, Past and Present* (Oxford: Pergamon Press).

Larbelestier, J. (1990) 'The Politics of Representation: Australian Aboriginal Women and Feminism', *Anthropological Forum*, 6: 2, 144.

Lowenthal, D. (1985) *The Past Is a Foreign Country* (Cambridge: Cambridge University Press).

Lozanovska, M. (1994) 'Abjection and Architecture: The Migrant House in Multicultural Australia', in Johnson (ed).

Mahoney, K., Malcolm, D., Hocking, B. and Threadgold, T. (1993) 'Introduction', *Australian Feminist Law Journal*, 1.

Mani, L. (1990) 'Multiple Mediations: Feminist Scholarship in the Age of Multinational Reception', *Feminist Review*, 35, Summer, 29.

Marguerite, R. and McNamara, (2000) Supplement to Chapter 8, in McRae et al.

Reynolds, H. (1996) *Frontier: Reports From the Edge of White Settlement* (Sydney: Allen and Unwin).

Rosaldo, R. (1989) 'Imperial Nostalgia', *Representations*, 26, Spring, 107–121.

Said, E. (1978) *Orientalism: Western Conceptions of the Orient* (London: Penguin).

Saunders, K. and Evans, R. (eds) *Gender Relations in Australia* (Sydney, Harcourt Brace and Janovich).

Shaw, W.S. (2000) 'Ways of Whiteness: Harlemising Sydney's Aboriginal Redfern', *Australian Geographical Studies*, 38, 3, 291–305.

Shaw, W.S. (2001) Ways of Whiteness: Negotiated Settlement Agendas in (Post)colonial Inner Sydney, unpublished PhD thesis, University of Melbourne.

Shaw, W.S (2003) 'Gendered Traditions and Aboriginalisms: Australian Courtroom Dialogue', *Gender Place and Culture*, 10, 4, 315—332.

Shaw, W.S (2005) 'Heritage and Gentrification: Remembering "the Good Old Days" in Postcolonial Sydney', in Atkinson R. and Bridge G. (eds) *The New Urban Colonialism: Gentrification in a Global Context*, London, Routledge, United Kingdom

Smith, N. (1996) *The New Urban Frontier: Gentrification and the Revanchist City* (London: Routledge).

Smith, N. and Williams, P.(eds) (1986) *Gentrification of the City* (Australia: Allen and Unwin).

Taylor, K. (1994) 'Things we Want to Keep: Discovering Australia's Cultural Heritage', in Hedon, Hooton and Horne (eds).

Thomas, N. (1994) *Colonialism's Culture: Anthropology, Travel and Government* (Cambridge: Polity).

Waitt, G. and McGuirk, P.M. (1996) 'Marking Time: Tourism and Heritage Representation at Millers Point, Sydney', *Australian Geographer*, 27, 1.

Ward, A. (1976) 'The Wholesome Precedent of W', *Legal Services Bulletin*, 2, 141–144.

Wasserman, R. (1984) 'Re-inventing the New World: Cooper and Alencar', *Comparative Literature*, 36, 2, Spring.

Chapter 8

A Work in Progress: Aboriginal People and Pastoral Cultural Heritage in Australia

Nicholas Gill and Alistair Paterson

Introduction

If you spend time in the region around Mistake Creek, an Aboriginal owned cattle station in the north western Northern Territory, you will soon run into Aboriginal people proudly sporting caps embroidered with the station name and logo. For these Aboriginal people, their association with a large cattle station is a source of pride and identity. This may be surprising to many, as a popular view is that the impacts of pastoral settlement have been unambiguously negative for Aboriginal people. Past and present Aboriginal associations with pastoralism are, however, diverse, encompassing everything from brutal violence to relatively benign, if paternal, labour relations, to contemporary Aboriginal ownership and management of pastoral enterprises and the not uncommon sight of Aboriginal cowboys in the inland and the north. Despite the relative economic decline of rural industries and critiques associated with environmentalism and the Aboriginal land rights movement, pastoralism maintains an influential position and high status in Australian society and continues to be celebrated in a variety of fora and through a range of events as an essential element of the economy and of Australian identity and mythology (Curthoys 2000; Gill 2005). Through their association with a station and through acquiring cattle working skills and the right to wear the trappings (hats, boots, belts, 'cowboy' shirts etc) associated with it, Aboriginal people are able to 'accrue the social and cultural capital that has historically rested with settler pastoralists' (Davis 2004, 39) For an account of how stockworkers earn such 'rights' see Strang (2001). This does not, however, imply that Aboriginal pastoralists have shed their Aboriginal identities and adopted an 'Aboriginality' that conforms neatly to settler pastoral identities and land use ideals. As Davis notes for the Kimberley in north western Australia, while practising pastoralism and accruing its benefits, Aboriginal pastoralists nonetheless maintain a 'radical alterity' (2004, 39) from settler pastoralists, partly through specific pastoral practices. Similar arguments have been made for Central Australia and northern Queensland by Gill (2005) and Smith (2003), who show that attempting to draw sharp boundaries between the pastoral and Aboriginal domains ignores their mutual production in both the past and in present daily life in which both change and stability are always evident.

This might seem a long way from a discussion of Aboriginal pastoral heritage, yet the heritages of pastoral Australia and of Aboriginal people have been, and largely still are, imagined and presented along lines that strongly demarcate the settler from the Aboriginal, the past from the present, and the material from worlds of interaction, values and associations (Byrne 1996; Harrison 2004). In this chapter we discuss recent approaches to heritage in Australia that have challenged this separation of the settler and Aboriginal pastoral domains and which are developing new perspectives on pastoral heritage that show the entangled and productive place of Aboriginal people in pastoral landscapes. Using case studies from pastoral areas of inland Australia (Figure 8.1) and from various time periods, we illustrate diverse associations of Aboriginal people with pastoralism, arguing that these are dynamic and creative in terms of the pastoral identities and forms of pastoralism that are emerging. These case studies include archaeological and geographical perspectives that find Aboriginal associations with pastoralism in and around the homesteads and woolsheds of pastoral stations; in memories of land, sites, and routes that embed former Aboriginal pastoral workers within pastoral landscapes; and through distinctively Aboriginal understandings of contemporary Aboriginal pastoral enterprises. We argue that attention to artefacts in the landscapes – buildings, fences, stockyards, campsite remains – if not fetishised or taken as natural signs of an 'unadulterated' history (Peet 1996) – in conjunction with diverse empirical sources can contribute to demystifying processes of landscape production and reproduction, producing detailed social histories of specific places (Rose and Lewis 1992).

Pastoral landscapes, national mythologies, bounded heritages

Despite pressure in recent times associated with the conservation and Aboriginal land rights and related movements, extensive pastoralism remains important in Australian national mythologies. The spread of extensive pastoral settlement and its (variable and arguable) success in surviving across large areas of the inland and the north, as well as in some areas of higher country in eastern Australia, is a key element to Australian frontier mythology and to national and local stories of settlers adapting to a new country and its vicissitudes of drought and floods (Gill 2005; Gill and Anderson 2005). This is not a story that goes without challenge in contemporary Australia (for example see Dominy 1997) and competing narratives of pastoral landscapes as either deathscapes or national heartlands have a long history in Australia (Haynes 1998; Heathcote 1987). Nonetheless the power and continued currency of Australian frontier mythology and its association with pastoralism can be seen in events such as high profile cattle droving re-enactments at moments of national celebration such as the 1988 Bicentennial and the 2002 'Year of the Outback' (Gill 2005). Less obviously, but more significantly, the debates over Aboriginal property rights on pastoral leases in the late 1990s and their political and legislative consequences led Howitt (2001) to argue that frontier thinking and the boundaries that it entails are deeply embedded and reproduced in many aspects of Australian life. The result is geographical practices characterised by an absence of the 'alien and incomprehensible Other' (Howitt 2001, 235), by refuge in the familiar and understood, and by socio-spatial uniformities and

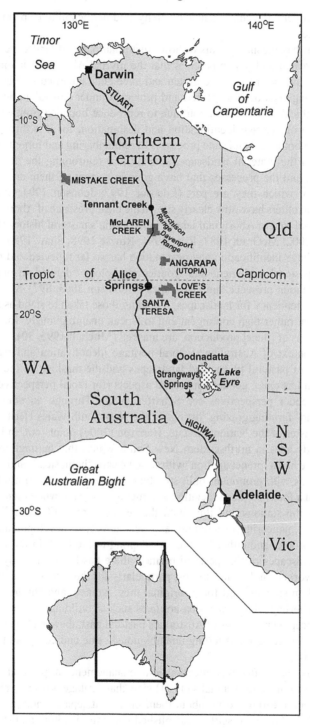

Figure 8.1   Locations referred to in the text

rigidities that deny the contingency, diversity, and permeability of peoples' lives in specific places.

What this has meant in terms of how Australian pastoral landscapes have been imagined, created and remembered is that the myriad Aboriginal involvements in pastoralism have been largely forgotten and have gone umarked insofar as pastoral cultural heritage has been identified and protected under Australian heritage laws. The geographical practices that continue to reproduce bounded frontier landscapes include those of heritage identification and designation, for it is through heritage practices that both local and state processes of remembering and forgetting find their expression in the material landscape, shaping and reinforcing the understandings about places and the processes that have gone into creating them and the broader landscapes of which they are part (Edensor 1997; Johnson 1994; 1999). While historians and others have now clearly shown the clear presence of Aboriginal people and the crucial role of Aboriginal labour in Australia's pastoral history (Cowlishaw 1999; Jebb 2002; McGrath 1987; Rose 1991; Rowse 1998; Shaw 1986), Australian pastoral heritage identification and designation has so far reflected and perpetuated a view of pastoral settlement and frontier mythology 'defined in opposition to Aboriginal labour, presence and landscapes' (Robinson 2005, 897). More generally this reflects a tendency for landscapes, including those taken to stand as heritage, to appear natural rather than contingent and to 'speak unambiguously' for themselves while the 'facts of [their] production' are masked (Mitchell 1996, 30).

In the context of Australian pastoral heritage identification and management, fetishisation of pastoral places and landscapes and the masking of the labour that created them has class, gender, and racial aspects (for racial perspectives see Rose 1992; for gender perspectives see Schaffer 1988). Through an enumeration of 'pastoral' and 'farming/grazing' listings in the New South Wales Heritage Register and the Register of the National Estate, Harrison (2004) points out, in his study of pastoral landscapes in north-eastern New South Wales, that heritage management agencies have had a 'preoccupation' with what he calls 'homesteads and woolsheds', in other words with prominent built structures (and other material artefacts). This has involved a focus on the association of structures with the white men who owned or managed the stations and thus with the 'great themes' of Australian pastoral settlement – pioneering, economic and technological development, subduing Aboriginal people, and either battling or adapting to the land. In the material and symbolic landscapes of this pastoral heritage there has been relatively little room for station workers and their families, particularly for the settler women who lived and worked on stations, and for Aboriginal men, women, and children. Within an approach to heritage that focuses on artefacts such as buildings as the embodiment of widely accepted narratives of Australia's pastoral past, these buildings are part of landscapes that are taken as telling unambiguously true and complete pastoral and settlement histories.

With this focus, the identification and management of pastoral heritage in Australia conforms to Lowenthal's (1994) view that heritage is usually about groups identifying and creating stories about themselves that separate and mark them out from others. In Australia, settlers established elaborate institutions through which they distinguished themselves and their European modes of settlement and land use from Aboriginal people who were viewed as less than human and, frequently, as objects of scientific curiosity, evidence of interesting but anachronistic cultures and physiology (Anderson 1998). Certainly, in European eyes, Aboriginal people did not improve and use land in ways that were equated with improvement and progress (Anderson 2003). Thus, from the early years of settlement and rural development, Aboriginal people were positioned as being of the past in contrast to the forward temporal location and movement of the settlers and their land use practices. Byrne has argued that Australian archaeology and Australian heritage regimes, themselves formatively and subsequently influenced by archaeology, have perpetuated this relative positioning of the Aboriginal people and the settlers (Byrne 1996; 2003). Archaeologists played key roles in framing and implementing state heritage laws and institutions from the 1960s in ways that reflected their own concerns with pre-contact Aboriginal sites and artefacts (Byrne 1996). Aboriginal heritage came to be institutionalised as that which related to pre-contact time and Aboriginality was thus both fixed and associated with 'deep time', the past, and 'tradition', rather than with history, the present, and the possibility of change and accommodation. In contrast, contact period heritage management has been almost exclusively focussed on settler activity or has artificially identified and delineated Aboriginal places as separate from settler places where in fact insertions and entanglements were present (Byrne 2003). This spatial separation occurs at multiple scales, from delineating white and black spaces within a town, to the notion that authentic Aboriginality is to be found only in the space and time of the outback. This separation in time and space of settlers and Aboriginal people reflects and reproduces an essentialism deeply embedded in Australian society (Head 2000). It exists in contemporary Australia, for example, in tourism marketing of Australia, in representations of outback areas, and in the attitudes and expectations of tourists themselves (Lane and Waitt 2001; Waitt 1999).

The material presented in this chapter is part of a shift in approaches to pastoral history and heritage. Rather than being informed by the separation, in time and in space, of settlers and Aborigines and restricted to a focus on 'sites', it focuses on relationships and entanglements between Aboriginal people and settlers and the places, networks, and landscapes that contained and were produced by these interactions. These are 'shared landscapes' (Harrison 2004) of pastoralism. In research and practice concerning such landscapes, there is an interest in the social and spatial details of how pastoral places were jointly occupied, the terms on which this occurred, and on the meaning of this joint occupation for people in these places today. There is less emphasis on buildings, artefacts and sites per se with a shift towards an emphasis on values and relationships between people and between people and places, and on heritage as a form of social action at the different scales through which heritage is created (Byrne et al. 2001; Harrison 2004; Johnson 1999). Consistent with this approach, in this chapter we emphasise the diversity of Aboriginal associations with pastoralism and pastoralists, the variety of Aboriginal pastoral landscapes, and the continued relevance and dynamism of Aboriginal associations with pastoralism, be they historical or contemporary.

Aboriginal labour and nineteenth-century pastoral settlement in the Lake Eyre Basin

In the far north of South Australia, in remote and arid country that today is part of one of Australia's largest inland cattle stations, can be found a range of buildings and other sites from one of the inland's earliest pastoral stations, Strangways Springs, founded in 1862. In keeping with the traditional focus of interest in pastoral history and heritage, the prominent stone buildings and other structures at the former station's headstation complex (homestead, workers' quarters, workshops) and woolshed areas are relatively well known to locals and 4WD tourists to the area. The site was added to the Register of the National Estate in 1995 and its statement of significance strongly reflects the tendency for pastoral heritage designation to focus on built structures and their association with themes such as the development of the pastoral industry (Department of Environment and Heritage 2006). Recent archaeological and historical research, however, has provided new perspectives on the station. This research has examined not only the headstation complex but also the broader structure of the station through identification and examination of the myriad outstation sites and camps across the station as well as biographical details of Aboriginal workers (Paterson 2003; 2005) In this case study we use both archaeological material and historical records to illustrate the role of Aboriginal people on the station and to show that, through material evidence such as that in pastoral-era camps and buildings, a story of interaction and shared places and landscapes can be told. To do this we will briefly discuss selected sites from Strangways Springs station, a wool scour and a shepherd's outstation, in conjunction with historical records, particularly letters written to the station owners by a manager in the 1860s. These sources can be studied to explore how the pastoral landscape was also that of many Aboriginal pastoral workers and their families as much as of the settlers.

The letters from pastoral managers from the earliest years of the station indicate the presence of Aboriginal workers. Despite this, the historical sources are biased towards reporting European rather than Aboriginal workers. As seen below with wool scouring, however, Aboriginal labour was essential to the station. For example, manager John Oastler referred to the need to recruit extra labour at peak times:

> [the] stock of flour, tea and sugar will not spin out…on account of the large number of Blacks [I] had to employ and consequently feed during the lambing season (Oastler, 23 June 1868)

An inexpensive workforce existed for the pastoralists whose dependence on Aboriginal labour is suggested in their letters and demonstrated at early pastoral sites by strong archaeological signatures of Aboriginal presence. Aboriginal work by both men and women was full-time (shepherding, bullock driving, translating, animal husbandry), seasonal (lambing and wool-washing) or temporary (message delivery, providing climatic and environmental information, and negotiating with Aboriginal people outside of the pastoral domain for seasonal labour and environmental information). Aboriginal women worked as shepherds at outstations and were also often given specialised tasks closer to the headstation, including care of sick animal and herding flocks of goats, fine-wool sheep, rams, and sheep intended for rations.

As with any large inland station, Strangways Springs is a network of related sites – headstation, outstations, and other work sites. They are places that were built by settlers and Aboriginal people. The sites were working and residential locales for family groups, as well as for white and black individual workers. The archaeological record of huts and campsites is complemented by assemblages of imported goods reflecting patterns of trade, commerce, and communication – ochre traded among Aboriginal people or ceramics brought from England. One site that was very important to the economic success of the station was a wool scour approximately one kilometre from the main headstation complex, probably used during the 1860s and 1870s (Figure 8.2). Pastoralists in dry regions such as Central Australia needed to wash, or 'scour' wool, in order to reduce its weight and thus its transport costs. For this task station managers needed a reliable work-force to wash the wool. There is little at the wool scour site to specifically suggest Aboriginal presence, for that we are dependent on the historical sources. The few remains of the wool scour comprised structural remains with concentrations of associated surface deposits, characterised by European debris. Structural remains suggest the presence of a waster race, water trough, hand-powered agitation and water reuse. Other built elements were a rectangular stone building and stone structural remains, possibly of fireplaces.

**Figure 8.2   Strangways Springs Head Station (right)
and Stockyards (left) looking west**

Photograph by Alistair Paterson.

Historical letters from Strangways reveal that Aboriginal people supplied much of the labour needed for washing wool. A letter written by the station manager in 1866, for example, illustrates the importance of Aboriginal wool-washers:

I can not keep the Blacks at the tubs – they can not stand the cold. I was depending on them as my chief stay. Last night a bitter disappointment occurred when I was getting on so well...our best washing Blacks have left us (Jeffreys, 27 May 1886)

These workers lived around the headstation complex and there are many contact campsites in its vicinity. These sites are characterised by clustering of settler cultural material such as food cans, razors, and buttons within scatters of quartz, glass and other fragments probably deriving from manufacture of tools by Aboriginal people. Emu shell and ochre are also present at at least one such site. There is very little structural material remaining and it is likely that people lived in simple bough structures similar to one photographed on the station in 1891 (Figure 8.3).

Figure 8.3   Image of Aboriginal campsite in Warriner Creek during the visit of the South Australian Pastoral Lands Commission in 1891

(Photograph: State Library of South Australia, PRG 280/1/40/121 reproduced with permission).

A second site, located in a dunefield on Strangways Springs is most likely one of a number of shepherds' out-stations built for the lambing during the 1860s. It comprises the remains of a hut, yards, and a nearby Aboriginal contact-period campsite (Figure 8.4). It represents an example of an historically-known element of early pastoralism that has rarely survived. Shepherds would stay at the outstation for up to several months, as long as water and grazing were sufficient for the sheep. The evidence from the shepherd's hut site strongly suggests cross-cultural contact use. Material derived from settler society included items such as buttons, cans, nails, wire, pipes, food and medicine bottles and jars and tools. Artefacts common to Aboriginal occupation were distributed in activity areas outside the hut (stone and glass tool reductions, ochre), inside the hut (ochre) and at the nearby Aboriginal camp. However, it is not clear whether Aboriginal residency was contemporary with European. This evidence at the site could equally result from both interracial interaction and avoidance. There is, however, evidence from station records that simultaneous use of outstations by settlers and Aboriginal people occurred. For example in 1866, a manager wrote that 'the same night the Blacks were camped...within six feet of the sheep. They had a carrobberry [corroboree] to announce themselves. Jones and King joined in it, I went to bed'.

Figure 8.4   Hut remains at Shepherding Outstation during excavation, with the roof poles *in situ*, 1995

Photograph by Alistair Paterson.

Twentieth century settlement and pastoral landscapes of labour and dwelling

In this section we want to focus not so much on the sites themselves, but on sites as places in the broader pastoral landscape, as alluded to in the Strangways Springs case. To do this we will discuss selected twentieth century pastoral sites that were important to an Aboriginal pastoral worker as nodes, that is 'single sites of investigation...through which disparate, but far reaching processes occur', processes linking individuals, social groups, and places in time and space (Nast 1996; Nast 2001, 74), Specifically we will consider two sites as part of the life, labour, movement, and memory of M. Kennedy. Kennedy was a Warumungu man who lived in and around the Murchison and Davenport Ranges near Tennant Creek in the Northern Territory

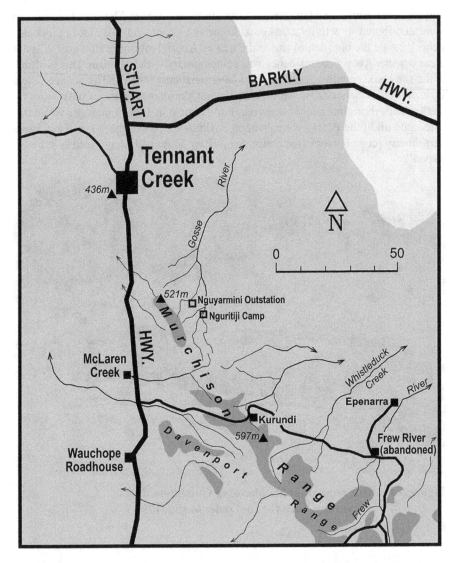

Figure 8.5   Murchison Davenport Ranges area with selected locations

and who worked on pastoral stations in the area from the 1930s to the 1970s. The two sites we will focus on are Nguyarmini and Nguritiji in relatively marginal pastoral country along the northern flanks of the Murchison Ranges (Figure 8.5). There are many places and routes in this area that were important in Kennedy's working life but these two places in particular appear in his memories and in archival documents. They appear not only in association with Kennedy's pastoral employment but also with a dynamic cast of characters including established settler pastoralists and a range of 'feral' settler and mixed descent pastoralists and their employees and wives, both settler and Aboriginal. The two sites feature strongly across different periods in various associations with pastoral activities at and around them, including a walkoff by Aboriginal workers from a nearby station in the 1970s.

By the 1920s pastoral settlement in the Murchison and Davenport ranges was consolidating along two broad lines. In poorer pastoral country in and around the north and west of the Murchison Ranges short term grazing licences were held by a series of small time operators, notably by the 1930s by the settler Sid Boon. Further east in the Murchisons and into and around the Davenport Ranges, a number of larger stations that remain today, such as Epanarra, Kurundi/Frew River, and Elkedra, were developing. Such heritage assessment that has been done in the region has mainly focussed on the homesteads and other structural remains at larger stations such as these at stock route bores (Pearce 1984). Although this is in keeping with traditional pastoral heritage assessment, it is also not surprising as these structures are relatively easy to identify and often have a strong archival presence. The pastoral landscapes, sites, and routes that we have explored with Kennedy (Gill 2005; Paterson et al. 2003) include more intangible aspects of Aboriginal associations with pastoralism and many of the widely scattered and often undocumented sites would be almost impossible to find without detailed local knowledge and their physical elements have often been largely or entirely destroyed by flood, fire, or termites.

Nguyarmini and Nguritiji are several kilometres apart on tributaries of the Gosse River and both places have histories that reach back prior to European settlement. Today they are both on Aboriginal land, immediately prior to which they were part of McLaren Creek cattle station or areas of vacant crown land. Prior to this they had been mainly covered by a series of grazing licences held by Sid Boon from the 1930s until his death in 1965. Both these places feature in Kennedy's life, from his earliest pastoral employment on Kurundi to his death in 2003. Over this entire period he maintained a strong association with pastoralism and cattle – from his stockman's clothing and his evident skills in butchering, yard construction, and rawhide and leather work, to his employment on stations across the region, his cattle droving trips around the NT and to Queensland and Western Australia, to more recent stints as manager of pastoral enterprises. In his earlier working life, Nguyarmini and Nguritiji were both places that Kennedy passed through on horseback as he travelled from stations further east, such as Kurundi, to places such as Kelly Well or Tennant Creek. When we knew Kennedy from the mid-1990s, Nguyarmini was an Aboriginal outstation and his home, and it was in the heart of his traditional country. Yet one of his first visits to this place was as a very young man in the company of George Birchmore, lessee of Kurundi station and one of Kennedy's first employers. This was probably in the mid to late 1930s and in this time of open range pastoralism, they had come on horseback from Kurundi to conduct a joint cattle muster of the Gosse River country with Sid Boon.

They came to Nguyarmini, then Boon's main home site, where they camped. They then mustered and camped through the Gosse River country where Boon had several cattle yards, two of which remain standing (Figure 8.6), and one of which at least Kennedy helped to build. There are also the remains of a goat yard at Boon's Nguyarmini camp and Kennedy recalled Aboriginal women working for Boon as goatherds. Kennedy's work with Boon included sorting and branding cattle at a yard made of piled mulga at Nguritiji, now destroyed by fire.

we bin come back to Nguritiji, old yard there. We bin brand'im right there. You know, we bin sort'im out all the Kurundi ones, tail'im out. And ol' Sid Boon cattle we bin give'im hand and brandem. Leave it right there (M. Kennedy, 17/6/2000).

Figure 8.6   Sid Boon's Nguyarmini cattle yards, 2000

Photograph by Nicholas Gill and Alistair Paterson.

Nguritiji was not only a pastoral site for Boon and a place where Kennedy worked for settler pastoralists (for more detail on Nguritiji see Paterson et al. 2003). More recently, Nguritiji had associations with pastoralism and Aboriginal people as the site where Kennedy, his brothers, his sister, and their families camped after they walked off Kurindi station in 1977 (Figure 8.7) as a result of long term dissatisfaction with working conditions and in anticipation of gaining ownership of some of their traditional land (Bell 1978). This largely spelled the end of their time working on stations and saw the beginning of an era where Aboriginal people in the area sought the return of their traditional lands:

> Just move off from Kurundi with the Toyota an' we 'ad the little car...Come back slowly through Panjiriji. An' keep, keep go back for get the tucker, flour, tea an' sugar, tobacco. Oh, get some meat from Station. No work, nothing. No work. Finish (M. Kennedy, 18/6/2000).

As Kennedy recalls, they continued an association with Kurundi, where they actually had good relations with the manager. Nguritiji eventually became an outstation with houses and associated infrastructure (Paterson et al. 2003). Kennedy and his brothers also kept horses and built a wooden horsebreaking yard which still largely stands. In the construction of this yard, for example in the design of the gate, is evidence of skills developed on cattle stations building yards using only basic tools and in the absence of materials such as hinges, bolts, and nails. There is also a post walkoff camp at Nguyarmini waterhole used by Kennedy and his siblings which was used before the outstation at Nguyarmini was built. Nothing, however, remains of this camp.

More recently Nguritiji has served as a temporary mustering camp for Kennedy's own pastoral enterprises, which have included both a commercial operation based at McLaren Creek and a small scale enterprise. These camps were largely made up of his sons and other younger male relatives. These activities have their own geography of camps, mustering locations, and routes for moving cattle, especially the route used when shifting Kennedy's share of the McLaren Creek cattle to his land when that enterprise ended. Kennedy's home, the Nguyarmini outstation, a few hundred metres from the waterhole and Boon's camp, has been the main base for this small scale pastoral enterprise.

Figure 8.7   Shelter at Nguritiji Kurundi Walk Off campsite, 2001

Photograph by Nicholas Gill and Alistair Paterson.

Contemporary Aboriginal pastoralism – meanings and motivation

It would be erroneous to assume that Aboriginal associations with pastoralism exist only in the past. A modest but rich body of ethnographic research examines recent and contemporary Aboriginal associations with pastoralism and the meaning of past associations in the present (Davis 2004; McGrath 1987; Minoru 2002; Rose 2004; Shaw 1986; Smith 2002). This work shows the importance of past and present associations with the pastoral industry to the identity of many Aboriginal people. In this section, we will use a case study of contemporary Aboriginal pastoralism in the Northern Territory to illustrate aspects of the importance of pastoralism for Aboriginal people. In particular, this section will show that the dynamic and mutually creative relationship between pastoralism and Aboriginal people is ongoing and that distinctive forms of Aboriginal pastoralism and pastoral landscapes are being created.

Since the 1970s, Aboriginal people in Australia have regained ownership of about one hundred pastoral properties, mainly in the inland and north (Phillpot 2001). In a study of commercial and non-commercial pastoral enterprises in the Northern Territory, Aboriginal pastoralists ranked the cultural and social benefits of pastoralism above other forms such as direct economic benefits (Gill 2005). Specifically, these Aboriginal pastoralists saw pastoralism having a role in the maintenance of Aboriginal culture. Pastoralism is now linked to Aboriginal values and society in ways that confound a singular interpretation of it as an economic or commercial activity. This section will draw on this research to examine these links.

Specifically, Aboriginal pastoralists saw pastoralism as a means by which to maintain Aboriginal culture and relationships to land. Customary Aboriginal ownership of country brings responsibilities as well as rights to the owners. This relationship with country is reciprocal and Aboriginal people must care for or 'hold' a country as one would carry a responsibility. 'Holding' denotes an active and intimate relationship between the holder and what is held (Myers 1986). An individual holds the country until the succeeding generation takes on the responsibility upon their death. The responsibilities of 'holders' to country include a range of activities such as protecting the country from damage, providing a new generation of owners educated in Aboriginal law to take over the responsibilities, and learning and performing the ceremonies that keep country and people strong and healthy (Rose 1992, 106–107).

Aboriginal pastoralists perceived that, despite having gained ownership of land, there remain future uncertainties in meeting obligations to country. Related to this, they worry that they are not fulfilling their obligations to the young people whom they 'hold' in a similar way to country. Concerns for young Aboriginal people and about the future of country are part of the desire to run cattle enterprises. In particular, these people were concerned about young people leaving their homes and the potential for excessive alcohol consumption and death and injury from violence or trauma such as car accidents; concerns that are well founded in disproportionately high Aboriginal mortality, morbidity, and injury rates in the NT (Territory Health Services 1998). The Aboriginal pastoralists were looking for ways to keep young people, especially the young men, on the country and so to prevent them from getting 'wild' and 'on the grog' (Aranda Aboriginal pastoralist). The problem of young people going into town and getting 'on the grog' is not only related to concern about their well-being and about fulfilling responsibilities to 'look after' them. As outlined earlier, in order to fulfil their obligations to country, owners of country have responsibility for providing a new generation of knowledgeable owners to succeed them. The possibility that the young people would not learn from them in time caused them to worry that they would fail to 'hold' the country. A Warumungu pastoralist expressed this fear of 'losing' the law and the country if the youngfellas are in town:

They lost the country, he might lost himself . . . he don't know anything about it, no ceremony business, he don't know sacred sites, he lost himself altogether. (Warumungu pastoralist)

In this context 'lost' has a particular meaning. Losing can be thought of as 'forgetting,' and as a significant cultural loss (Arthur 1996). Myers (1986), however, indicates that, in relation to country, the concept carries the implication of handing that country on, of losing it on death, but leaving it for the next generation. These men just quoted appear to fear a more serious loss; the loss of the country not only to themselves upon their death but also to their children, who they fear may not equipped to take up responsibility for country. Rights to country must be maintained by visiting country and sites, learning the stories and rituals, and keeping country and sites 'clean'. Instruction in these matters by older men is a 'crucial component of the social reproduction of ownership and through it the production of adult men' (Myers 1986, 151), Cattle work was seen as a means by which young men could be enticed to be 'on the country' in the company of knowledgeable older men. Why do these Aboriginal pastoralists see a role for pastoralism in helping to ensure cultural continuity? The answer lies in the role cattle work played in their own lives and in 'growing them up' and in the ways in which Aboriginal and cattle cultures complemented each other. Both McGrath (1987) and Baker (1999) have observed that the activities of Northern Territory cattle work, checking waterholes, checking pastures, mustering cattle, and working in stock camps, doubled as opportunities to learn about country and to fulfil obligations to country. Such moments of gaining knowledge of country are evident in the recollections of cattle work among Alyawarra and Warumungu pastoralists. When these men were working on stations, their work provided them with time on country as they rode over it checking on cattle and waters. Travelling and talking with a Warumungu Aboriginal pastoralist in particular revealed his intricate geography of travel routes, waterholes, and sites in the region that he used for various aspects of his pastoral work. As a result, he is now able to demonstrate ownership and he needs to pass this ability on to his sons:

My father's country. So I got to follow that. And all our sons . . . People getting old and old. He's the one that got to come along, second, to look after country. We used to shift 'em cattle and bring horses. You got fill up your canteen...Traditional owner, people belonging to country, well he know all the rockholes . . . that where people got to be, – to look after place you know. Keep up with the country so long as young fellas stick to daddy. (Warumungu pastoralist)

For these older men, cattle work was part of the process, the 'proper way', by which Aboriginal customary ownership was reproduced, and by which they were made into men who had the 'qualities and discipline associated with adulthood' (McGrath 1987, 167) particularly the ability to meet their responsibilities in Aboriginal law. By instilling in their young men the ability to do cattle work and thereby keeping them on their country they hope to secure the future of people and country by practicing 'two ways':

We should cut 'em from them youngfellas [the drinkers] . . .We got to teach'em all that one . . . cattle way and business (ceremony) way . . . so they can understand two way. (Alyawarra pastoralist)

These Aboriginal pastoralists perceive that security of 'ownership' comes through mastering both Aboriginal and non-Aboriginal ways. They see pastoralism as a means by which this can be achieved and as an activity that can assist in reproducing Aboriginal ritual life and land ownership. For these Aboriginal pastoralists, the relevance of their working lives on pastoral stations is not its contribution to an Australian pastoral heritage, but its role in contemporary Aboriginal identities and in tackling present day social issues facing Aboriginal people.

Conclusion

Byrne (2003) has argued that a focus on sites has been 'debilitating' for Aboriginal cultural heritage management. A focus on sites without attention to their spatial, temporal, and social context enacts 'spatial containment' that takes 'Aboriginal contact experience out of the larger colonial landscape and confine[s] it to places where white people rarely went' (Byrne 2003, 188). For pastoral heritage management under the 'homesteads and woolsheds' model, a similar process has occurred in relation to pastoral heritage identification and management (Harrison 2004). In both cases there is a 'static conception of space and the past is fossilised 'through the reproduction of material culture' reducing 'the dynamism of historical processes' (Johnson 1999). As these authors have also argued, however, seeing sites as places of interaction and intersection, both internally and beyond allows for the recovery of meaning at those sites and for attention to be paid to their complex and temporally variable roles and dynamics such that histories and landscapes *founded* on social separation and spatial differentiation become untenable. In our discussion of Aboriginal associations with pastoralism we have used sites as a way into detailed histories of pastoral place and landscapes. This has clearly illustrated the complex interactions between settlers and Aboriginal people that occurred at a range of different kinds of pastoral sites and across pastoral landscapes. Our discussion of two sites important to M. Kennedy showed that a static association of those sites with pastoralism is inadequate in understanding the range of different types of associations with pastoralism and different types of people evident at any one site over time. Moreover, Kennedy's pastoral landscape also showed that the significance of particular sites derives from their linkages to other places and the movement that occurred in the course of pastoral work. Collectively, our three cases also demonstrate the great diversity of Aboriginal associations with pastoralism and their ongoing construction as Aboriginal pastoral enterprises and contemporary Aboriginal pastoral identities develop.

As Harrison (2004) notes, a conception of pastoral heritage based in landscapes and a range of intangible associations, values, networks, and ephemeral sites and which sees these landscapes as shared and active rather than as static texts by which to tell a story of settler triumphalism, poses challenges for the 'homesteads and woolsheds' model. Pastoral heritage in Australia is far more complex and spatially variable than this model implies and indeed, it is also more complex than the very narratives about settlement, race and progress upon which it is based, have allowed for. It is insufficient for Aboriginal associations to be simply tacked on to existing notions of pastoral heritage; to do would simply reproduce the separations that already exist. Pastoral heritage requires fundamental reordering. A key way forward will be through attention to historical processes, to the range of actors, and to spatial dynamism rather than to stasis at the local level and through the illustration of how these factors relate to larger scale processes without identifying and putting these first as has largely occurred to date. By taking this approach Australian pastoral heritage will be enriched and we will generate far more compelling and diverse stories about land and people in Australia's pastoral landscapes for the wider audiences who visit such places.

References

Anderson, K. (1998), 'Science and the Savage: The Linnean Society of New South Wales, 1874–1900', *Ecumene:* 5:2, 125–143.

—— (2003), 'White Natures: Sydney's Royal Agricultural Show in Post-Humanist Perspective', *Transactions of the Institute of British Geographers* 28, 422–441.

Arthur, J. M. (1996), *Aboriginal English: A Cultural Study* (Melbourne: Oxford University Press).

Baker, R. (1999), *Land Is Life: From Bush to Town – The Story of the Yanyuwa People* (Sydney: Allen and Unwin).

Bell, D. (1978), 'For Our Families: The Kurundi Walkoff and the Ngurrantji Venture', *Aboriginal History* 2:1–2, 32–62.

Byrne, D. (1996), 'Deep Nation: Australia's Acquisition of an Indigenous Past', *Aboriginal History* 20, 82–107.

—— (2003), 'The Ethos of Return: Erasure and Reinstatement of Aboriginal Visibility in the Australian Historical Landscape', *Historical Archaeology* 37:1, 73–86.

—— (2003), 'Nervous Landscapes: Race and Space in Australia', *Journal of Social Archaeology* 3:2, 169–193.

Byrne, D., et al. (2001), *Social Significance: A Discussion Paper* (Sydney: NSW National Parks and Wildlife Service).

Cowlishaw, G. (1999), *Rednecks, Eggheads and Blackfellas: A Study of Racial Power and Intimacy in Australia* (Sydney: Allen and Unwin).

Curthoys, A. (2000), 'Mythologies', in Nile R. (ed.) *The Australian Legend and Its Discontents* (Brisbane: University of Queensland Press).

Davis, R. (2004), 'Aboriginal Managers as Blackfellas or Whitefellas? Aspects of Australian Aboriginal Cattle Ownership in the Kimberley', *Anthropological Forum* 14:1, 23–42.

Department of Environment and Heritage '*Australian Heritage Database*' <http://www.deh.gov.au/heritage>, accessed 13/6/2006.

Dominy, M. D. (1997), 'The Alpine Landscape in Australia: Mythologies of Ecology and Nation', in B. Ching et al. (eds) *Knowing Your Place: Rural Identity and Cultural Hierarchy* (New York: Routledge).

Edensor, T. (1997), 'National Identity and the Politics of Memory: Remembering Bruce and Wallace in Symbolic Space', *Environment and Planning D: Society and Space* 29, 175–194.

Gill, N. (2005), 'Aboriginal Pastoralism, Social Embeddedness and Cultural Continuity in Central Australia', *Society and Natural Resources* 18, 699–714.

—— (2005), 'Life and Death in Australian 'Heartlands': Pastoralism, Ecology and Rethinking the Outback', *Journal of Rural Studies* 21, 39–53.

Gill, N., et al. (2005), 'Improvement in the Inland: Culture and Nature in the Australian Rangelands', *Australian Humanities Review* 34.

Gill, N., Paterson, A. and Kennedy, M. (2005), '"Murphy, Do You Want to Delete This?" Hidden Histories and Hidden Landscapes in the Murchison and Davenport Ranges, Northern Territory, Australia', in G. Ward et al (eds.) *The Power of Knowledge, the Resonance of Tradition* (Canberra: Aboriginal Studies Press).

Harrison, R. (2004), *Shared Landscapes: Archaeologies of Attachment and the Pastoral Industry in New South Wales* (Sydney: University of NSW Press and Department of Environment and Conservation).

Haynes, R. D. (1998), *Seeking the Centre: The Australian Desert in Literature, Art and Film* (Cambridge: Cambridge University Press).

Head, L. (2000), *Second Nature: The History and Implications of Australia as Aboriginal Landscape* (Syracuse: Syracuse University Press).

Heathcote, R. L. (1987), 'Images of a Desert? Perceptions of Arid Australia', *Australian Geographical Studies* 25, 3–25.

Howitt, R. (2001), 'Frontiers, Borders and Edges: Liminal Challenges the Hegemony of Exclusion', *Australian Geographical Studies* 39:2, 233–245.

Jebb, M. A. (2002), *Blood, Sweat and Welfare: A History of White Bosses and Aboriginal Pastoral Workers* (Nedlands: University of Western Australia Press).

Johnson, N. (1994), 'Cast in Stone: Monuments, Geography, and Nationalism', *Environment and Planning D: Society and Space* 13, 51–65.

—— (1999), 'Framing the Past: Time, Space and the Politics of Heritage Tourism in Ireland', *Political Geography* 18, 187–207.

Lane, R., et al. (2001), 'Authenticity in Tourism and Native Title: Place, Time and Spatial Politics in the East Kimberley', *Social and Cultural Geography* 2:4, 381–405.

Lowenthal, D. (1994), 'Identity, Heritage and History', in R. Gillis (ed.) *Commemorations: The Politics of National Identity* (New Jersey: Princeton University Press).

McGrath, A. (1987), *Born in the Cattle: Aborigines in Cattle Country* (Sydney: Allen and Unwin).

Minoru, H. (2002), 'Reading Oral Histories from the Pastoral Frontier: A Critical Revision', *Journal of Australian Studies* 72, 21–28.

Mitchell, D. (1996), *The Lie of the Land: Migrant Workers and the Californian Landscape* (Minneapolis: University of Minneapolis Press).

Myers, F. R. (1986), *Pintupi Country, Pintupi Self: Sentiment, Place and Politics among Western Desert Aborigines* (Washington and Canberra: Smithsonian Institution Press and Australian Institute of Aboriginal Studies).

Nast, H. (1996), 'Islam, Gender and Slavery in West Africa circa 1500: A Spatial Archaeology of the Kano Palace, Northern Nigeria', *Annals of the Association of American Geographers* 86:1, 44–77.

—— (2001), 'Nodal Thinking', *Historical Geography* 29, 74–76.

Paterson, A. (2003), 'The Texture of Agency: An Example of Culture-Contact in Central Australia', *Archaeology in Oceania* 38, 52–65.

—— (2005), 'Early Pastoral Landscapes and Culture Contact in Central Australia', *Historical Archaeology* 39, 28–48.

Paterson, A., et al. (2003), 'Archaeology of Historical Realities? Two Case Studies of the Short Term', *Australian Archaeology* 57, 82–89.

Pearce, H. (1984), *Tennant Creek Historical Sites Survey – Report to the National Trust of Australia (N.T.)* (Darwin: National Trust of Australia (N.T.)).

Peet, R. (1996), 'A Sign Taken for History: Daniel Shay's Memorial in Petersham, Massachusetts', *Annals of the Association of American Geographers* 86:1, 21–43.

Phillpot, S. (2001), 'Understanding Whitefella Secret Cattle Business', in R. Baker et al (eds.) *Working on Country: Contemporary Indigenous Management of Australia's Lands and Coastal Regions* (Melbourne: Oxford University Press).

Robinson, C. (2005), 'Buffalo Hunting and the Feral Frontier of Australia's Northern Territory', *Social and Cultural Geography* 6:5/6, 885–901.

Rose, D. B. (1991), *Hidden Histories: Black Stories from Victoria River Downs, Humbert River and Wave Hill Stations* (Canberra: Aboriginal Studies Press).

—— (1992), *Dingo Makes Us Human: Land and Life in an Aboriginal Australian Culture* (Cambridge: Cambridge University Press).

—— (1992), 'Nature and Gender in Outback Australia', *History and Anthropology* 5:3–4, 403–425.

—— (2004), *Reports from a Wild Country: Ethics for Decolonisation* (Sydney: University of NSW Press).

Rose, D. B., et al. (1992), 'A Bridge and a Pinch', *Public History Review* 1, 26–36.

Rowse, T. (1998), *White Flour, White Power: From Rations to Citizenship in Central Australia* (Melbourne: Cambridge University Press).

Schaffer, K. (1988), *Women and the Bush: Forces of Desire in the Australian Cultural Tradition* (Cambridge: Cambridge University Press).

Shaw, B. (1986), *Countrymen: The Life Histories of Four Aboriginal Men as Told to Bruce Shaw* (Canberra: Australian Institute of Aboriginal Studies).

Smith, B. R. (2002), 'Pastoralism, Land and Aboriginal Existence in Central Cape York Peninsula', *Anthropology in Action* 9:1, 21–30.

—— (2003), 'Whither "Certainty"? Coexistence, Change and Land Rights in Northern Queensland', *Anthropological Forum* 13:1, 27–48.

Strang, V. (2001), 'Of Human Bondage: The Breaking in of Stockmen in Northern Australia', *Oceania* 72:1, 53–78.

Territory Health Services (1998), *The Aboriginal Public Health Strategy and Implementation Guide 1997 – 2002* (Darwin: Territory Health Services).

Waitt, G. (1999), 'Naturalising the "Primitive": A Critique of Marketing Australia's Indigenous People as "Hunter-Gatherers"', *Tourism Geographies* 1:2, 142–163.

Chapter 9

Lobethal the *Valley of Praise*: Inventing Tradition for the Purposes of Place Making in Rural South Australia

Matthew W. Rofe and Hilary P.M. Winchester

Introduction

The village of Lobethal, in the Adelaide Hills, was established by German Lutherans in 1842 and so has a long and distinctive Germanic heritage. Located some 35 km northeast of Adelaide in the Mount Barker ranges, Lobethal is an idyllic rural settlement surrounded by orchards, with a tranquil and wholesome ambience far removed from the reach of the sprawling metropolis of Adelaide. It is well known in South Australia for its Christmas lights which draw over 250,000 visitors each year. Its rural character and association with Christmas traditions conceal a troubled history characterised by periods of conflict, struggle and turmoil that problematise the commonly perceived tranquillity of rural areas.

The aim of this chapter is two-fold. First, it traces the changing nature and representations of Germanic heritage attributed to the place referred to, since European colonisation, as Lobethal.[1] This history reveals the complex, changing and contested nature of local place identity, and demonstrates the ways in which international events are played out in different localities. Second, the chapter turns to a critical investigation of Lobethal's contemporary construction as a Christmas wonderland as a specific form of rural idyll place making. The seventeen day Lights of Lobethal festival draws upon a carefully and deliberately constructed intersection between the village's Germanic Lutheran heritage and romanticised notions of rural community (Winchester and Rofe 2005). Central to the Festival is the mobilisation of sentiments that equate rurality with tranquillity, simplicity and 'old fashioned' values. In short, Lobethal's Christmas wonderland is a social construction that strategically utilises discourses of the rural, the religious and the

---

1 While this chapter does not explicitly deal with issues of Indigenous occupation of the area under discussion prior to European settlement, it acknowledges the Peramangk as the traditional owners. Further, the authors wish to acknowledge the devastating impact colonisation had upon the Indigenous groups of the land now referred to as South Australia, especially the Peramangk whose culture was destroyed. Today, little is known about the Peramangk. That knowledge which does survive is being reconstructed from the oral traditions of other Indigenous groups, most notably the Kuarna of the Adelaide Plains. For further information on Indigenous dispossession in South Australia see Foster *et al.* (2001).

community. This critical examination of the changing representations of Lobethal's identity and landscape provides the opportunity to unravel the complexity inherent in the landscape, the highly subjective nature of heritage place making and the commodification of the rural idyll discourse.

Rural studies have enjoyed a long and fruitful tradition within the social sciences. Once considered as ideal study sites due to their perceived discretely bounded geography, the rural came to be considered the domain of 'natural' communities because of its perceived isolation and self-containment (Frankenberg 1975; Stacey 1969). However, more recent studies have uncovered the socially constructed nature of rural landscapes and their communities (Sibley 1995; Matthews *et al.* 2000). A central theme of such studies has been to trace the emergence of the discourse of the rural idyll or rurality. Critical rural studies, employing deconstructive forms of analysis, have exploded simplistic notions of the rural as 'sleepy' and harmonious. In this vein, Newby (1987, 1) demanded that '[t]he conventions which surround a romantic view of the countryside... need to be cleared aside...' along with the '...equally pervasive evaluation of rural life... that nothing of importance ever happens there: that Arcadian virtues exist beneath a pall of tedium'. Critical rural studies have opened the floodgate for more innovative work problematising the rural idyll. The rural landscapes examined in such studies are riven by difference, conflict and exclusion (Mingay 1989; Sibley 1995; Little and Austin 1996; Cloke and Little 1997; Matthews *et al.* 2000). Despite this recognition of conflict and complexity, the rural idyll remains a powerful discourse. These varied understandings of rural space provide a framework for understanding the changing representations of Lobethal over time.

Rurality and the romantic

The rural as a 'place apart' is deeply embedded in the Australian psyche. The origins of this romanticised rural sentiment can be traced to the rapid onset of urbanisation during the British Industrial Revolution (Bunce 1994; Bessière 1998; Hopkins 1998) long before the establishment of the Colony of South Australia. Industrial cities were vilified as '...unnatural embossments... odious wens, produced by corruption and engendering crime, misery and slavery' (Cobbett 1912, 43, cited in Bunce 1994, 14). The urban squalor and resultant social disarray gave rise to longings for a purer form of existence. This longing found its muse in idyllic notions of the rural. Contrasting the 'odious' industrial city, the rural was construed as '...a place of 'community', where innocence, safety, friendship and family values still prevail[ed]' (Hopkins 1998, 346). Unsurprisingly, with increasing urbanisation in both Europe and Australia, the rural came to be viewed with a sense of nostalgia. Little and Austin (1996) argue that the construction of the rural ideal embodies an 'enduring myth'. As a form of myth the rural idyll constitutes a symbolic foil providing an idealised, alternate version of our urban reality. As Short (1991, 34) argues the rural is:

>...often used in contrast with the fears of the present and the dread of the future...
>Households can look back to rural roots... the location of nostalgia, the setting for the

simpler lives of our forebears, a people whose existence seems idyllic because they are unencumbered with the immense task of living in the present.

The stability of the rural idyll has stimulated Hopkins (1998, 77) to argue that the term 'rural' has itself emerged as a '... brand name for a specific kind of place commodity;... a symbolic countryside'. Rural branding occurs both as the commodification of places and the packaging of events. A significant body of work has emerged examining the multitude of rural festivals that commodify, amongst other forms of rurality, cuisine (see for example Bessière 1998) and folk festivals (see for example Smith 1993; Halewood and Hannam 2001). Essentially, the rural brand packages a place-specific item, event or landscape to be consumed.

Places can be commodified using the rural brand and by manipulating the discourses of the rural idyll for economic benefit (see Bessière 1998; Ekman 1999; Hansen 1999; Tonts and Grieve 2002; Panelli *et al.* 2003). Deliberate place marketing strategies allow '[r]ural populations [to extend] their networks, widening their social space and economic scope' (Bessière 1998, 22). This widening of rural space and scope has been fuelled by economic decline within the agricultural sectors of many western nations. Thus, rural communities '...have [been forced to] become more conscious of their own image and of the importance of local culture' (Ekman 1999, 282). The rural idyll provides an established discourse easing the transition from agricultural production to 'rural' consumption for some rural communities. This 'scene change' (Panelli *et al.* 2003, 390) provides an opportunity to examine rural places as socially constructed and mediated landscapes.

Geographers have long recognised the socially constructed and mediated nature of landscape. Lewis (1979, 12) encapsulated this recognition by asserting '...*all human landscape has cultural meaning*' (original emphasis). Similarly, Hopkins (1998, 79) asserts that '[i]magination is... the place where our landscapes begin'. The meanings attached to specific places are negotiated by social agency and at times through social conflict. Traditionally, landscape analyses have provided richly textualised accounts of urban environments, particularly those experiencing the changes associated with place marketing (see for example Paddison 1993; Dunn *et al.* 1995; Rofe 2004). However, as Panelli *et al.* (2003) lament, rural studies have largely been devoid of a critical landscape analysis perspective. Landscapes here denote a richly symbolic text. The human imagination is of course selective in both its application and its recollection. The idyllic imagination of the rural draws on a nostalgic and romanticised view, which is partial, commodifiable and manipulable. The notions of social harmony, community and safety that are central to the rural idyll are increasingly being problematised. Recent studies have revealed the rural as a site of alienation, oppression and exclusion (Jackson 1989; Mingay 1989; Sibley 1995; Cloke and Little 1997). Exemplifying this growing literature, Newby's (1987, 127–137) history of rural unionism in Great Britain during the late 1880s depicts a landscape of unrest and violence.

Lobethal's short history is equally tumultuous. Despite the carefully constructed veneer of tranquillity and harmony, Lobethal has experienced episodes of religious conflict, racial exclusion and economic depression. These alternate rural landscapes are neither idyllic nor marketable. Thus, they often remain invisible or are deliberately suppressed by powerful interest groups lest they disturb the harmonious rural idyll. Yet, they are undeniably parts of the rich tapestry that is Lobethal's landscape. It is to these aspects of Lobethal's heritage that we now turn.

The Germanic villages: religious havens or seditious hideouts?

Pioneering settlement – religious haven?

South Australia has a long and proud tradition of German settlement. Fleeing religious persecution, some 2,500 German Lutherans immigrated to the colony of South Australia between 1837 and 1841. Under the stewardship of Pastor August Kavel and with the financial assistance of George Fife Angus, a founding member of the South Australian Company, the newly arrived exiles quickly began to establish a number of villages on the rural fringes of colonial Adelaide. First establishing the village of Klemzig, some 5 km to the north of Adelaide, the new migrants quickly settled the Adelaide Hills region establishing the village of Hahndorf in 1839 and Lobethal in 1842. The isolation, cultural and religious homogeneity and kinship bonds of these villages resulted in the development of a strong sense of community and place-specific identity. Indeed, the Germanic villages developed a unique sense of 'apartness' from the sprawling city of Adelaide despite their relative proximity. The replication of Germanic cultural traditions, dress and building styles in the Adelaide Hills created a distinctive community within the British colony. The Germanic villages quickly came to be romanticised as idyllic places within the harsh Australian landscape. One observer in 1840 wrote that in Adelaide '[t]he weather was so hot it was almost insupportable and not a blade of grass…' grew, yet in the Germanic villages '…the air felt so pure and invigorating that I could not think that I was in the same country as Adelaide' (cited in Whitelock 2000, 360). Such depictions are the genesis of the Germanic Lutheran villages as idyllic rural retreats. According to Bunce (1994), these images were deeply rooted in the psyche of the Victorian period. Contrasting the urban squalor spawned by the industrial revolution, rural landscapes were romanticised as places of simplicity, community and purity. Epitomising these sentiments, one English visitor to the Adelaide Hills noted that in the Germanic villages:

> Scarcely a day passes but that some of the people repair to the place of worship, wither early in the morning, or in the evening after work, for the purpose of returning thanksgiving to the Supreme Ruler and Protector, in this foreign land, a truly excellent custom, and in which lies one of the elements of their prosperity… All is simplicity and harmony (cited in Brauer 1985, 43).

The perceived pious nature of the Germanic Lutherans and their efforts in rapidly expanding the colony, earned them their reputation as hardy and industrious pioneers. In effect, the German Lutherans were considered as exemplars of the correct moral fibre so desired by the architects of the free colony. Unlike other Australian colonies, South Australia was meticulously planned as a settlement free from convicts and undesirable elements so as to '…not place a scattered and half-barbarous colony on the coast of New Holland, but to establish... a wealthy, civilized society' (Wakefield cited in Whitelock 2000, 3). Under this vision, the German Lutherans were perceived as the perfect immigrants. Commenting upon their positive qualities, Francis Dutton (1846, 161) observed:

> Now see how different the German labourer in the colony acts: the necessity of every farthing he spends, is seriously weighed, before he parts with it, you never see a German in a public house drinking spirits; he will come into the town many miles afoot, carrying, perhaps, a heavy load of vegetables, or what not, for the market; after he has sold his goods, he will take a lump of bread out of his pocket, brought with him from home, of his wife's own baking, and his day's profit must have been very good to induce him to buy even a glass of ale to wash down his frugal dinner; more frequently it is a draught of spring water.

This excerpt captures both the serious and wholesome nature of the Germanic Lutherans. However, the industrious German settlers and the tranquillity of their villages were little more than a veneer that masked harsher and more problematic realities.

The upheaval associated with and legacy stemming from their flight from Prussia cannot be underestimated. Rather than simple peasant folk seeking a new land of religious tolerance, the German Lutherans were a forcibly displaced group having faced severe repression for resisting King William III's merger of the Reformed and Lutheran Churches (see Gerber 1984). The Lutherans were '...exiles – refugees less pulled to the countries of destination than pushed away from their place of origin' (Gerber 1984, 500). State repression took many forms, including imprisonment and confiscation of property. In many instances, those who decided to come to Australia had no means of paying for their or their families' passage. While seeking assistance in England, Kavel encountered George Fife Angus, who agreed to provide financial assistance by chartering the *Prince George* for their journey. Angus' actions have been cast as those of a great philanthropist and humanitarian. Indeed, a grand monument to Angus as a 'Patriot, Politician and Philanthropist' stands on the banks of the River Torrens in Adelaide. This monument, erected by the Angus family in 1915, bears a large bronze plague depicting the German Lutherans boarding the *Prince George*. However, another reading of the motivations of Angus and the conditions under which the Germanic Lutherans toiled is possible. This counter-reading problematises the romanticised notions of this group and their contribution to the colony.

Beyond the rhetoric of philanthropy, the strategic use of the Germanic Lutherans to develop Angus' personal land-holdings and invigorate the progress and economy of the floundering colony is evident. Indebted to Angus, for both the assisted passage and land purchases from his business holdings upon arrival, the immigrants found themselves working for many years to repay their debts. The weight of this hung particularly heavy upon Pastor Kavel (cited in Brauer 1985, 34), who in a letter to Angus wrote:

> I admire your generosity and Christian love, and I pray that the Lord may reward you. Your kindness has put a heavy burden of obligation on my heart, for have I not considered all those advances made to my people as if they had been made to myself?.. I pray, my

dear Sir and friend, that the Lord may never suffer me to forget it, but enable me to show my gratitude to the very end of my life.

The repayment of Angus' loans precipitated changes to the tradition of wedding dowries with the emergence of grooms assuming their brides' and in some cases her family's, 'ship debt' (see Gerber 1984; Brauer 1985). Despite his sentiments of Christian brotherhood, Angus charged excessive rates of interest upon his philanthropic advances. Brauer (1985, 36) estimates that Angus was making profits of up to 400 per cent on sales of land and provisions to the German Lutherans.

Under such conditions, the establishment of the Germanic Lutheran villages was extremely arduous. Accounts exist, detailing how Germanic communities in the Adelaide Hills ate lizards, grasses and roots during poor harvest periods when surplus produce from meeting repayment conditions were lean. These accounts call into question the sense of vigour and health that external accounts of the Germanic villages, such as those cited above, created and perpetuated. Woodruff's (1984) medical history of South Australian colonisation undermines these accounts further, detailing the occurrence of a range of health problems amongst the Germanic villages. The most serious of these was a series of typhoid fever outbreaks during the 1880s, many of which occurred solely in the Germanic Villages.

While the tribulations of debt, environmental hardship and pestilence were external forces against which the Germanic Lutherans triumphed, internal trials were evident within this community. These internal trials further problematise the solidarity of kinship and the bonds of faith so often attributed to the Germanic Lutherans. Rather than being religiously and socially harmonious, schisms on matters of faith and social standing split the group on a number of occasions. Most notable amongst these was the emergence of religious differences in the village of Hahndorf that prompted 18 families to leave and establish the village of Lobethal in 1842. Nor was the Lobethal community immune from religious disharmony. Indeed three religious splits occurred in Lobethal between 1846 and 1876 resulting in the establishment of four separate churches (see Young *et al.* 1983, 16–17). This turmoil created a distinct social stratification within the village, further fuelled by the arrival of wealthier settlers (Young *et al.* 1983, 20). Purchasing large tracts of land for agricultural and industrial purposes, the new arrivals marked the emergence of Lobethal as a diversifying and significant economic region within the Adelaide Hills. Epitomising this was the establishment of the Kleinschmidt Brewery and the Lobethal Tweed Factory. Religious fractures and social turmoil, accompanied by the emergence of industrial enterprises within Lobethal problematise the romantic construction of the village as a harmonious rural haven.

Wartime tensions – seditious hideout?

Idyllic perceptions of the Germanic villages were dealt a further blow with the commencement of World War I. Despite having sworn an oath of allegiance to Queen Victoria upon arriving in the colony, German immigrants were demonised as potential agents provocateurs. Fuelled by xenophobic hysteria, the House of Assembly established a Nomenclature Committee to investigate the 'disturbing' number of Germanic place names within South Australia. Mobilising the rhetoric of patriotism, the House declared on 2 August 1916 that:

> The time has now arrived when the names of all towns and districts in South Australia which indicate foreign enemy origin should be altered, and that such places should be designated by names either of British origin or South Australian native origin (cited in Whitelock 2000, 364).

The Nomenclature Committee expunged 69 Germanic town names from the map (Australian Bureau of Statistics 2002). This action was met with some community opposition. One article in *The Register* (5 August 1916, 13), Adelaide's largest newspaper of the time, lamented that '[r]easons may be given why some of them [German place names] should be retained. There are elements of romance and of historical fitness in certain appellations'. However, this article concluded that '...the action... taken is... a mild reprisal for the unnameable brutalities committed by Germany on land and sea. It may be unfortunate, but it is patriotic!' Symbolically linking place names with a propaganda discourse of atrocities created a powerful discourse repudiating South Australia's Germanic heritage. Emphasising this, another article from *The Register* (3 August, 1916) declared:

> Today the German names are red with memories which tear women's hearts and make strong men weep. They are wrapped in an atmosphere of poisonous gas, and drip with the blood of heroes and heroines. Britons instinctively turn from them as from repulsive scenes.

Lobethal's name was changed to Tweedvale, an Anglicised reference to the Tweed factory, in 1917. Corresponding with this name change the Lutheran day school was shut, as were all German newspapers. Adding further insult, numerous Germanic place names were replaced with the names of either British Generals or victorious battlefields. Thus Kaiser Stuhl was renamed Mount Kitchener and Gruenthal became Verdun. Under such xenophobic weight many families anglicised their names. The name Lobethal was not reinstated until December 1935 as a 'Centenary gesture' (Young *et al.* 1983, 17) while others were reinstated even later.

External prejudice and the erasing of the town's Germanic heritage notwithstanding, Lobethal prospered economically during the First World War and the Great Depression. The town's population had steadily increased (1881–220 people; 1911–731 people; 1933–1219 people), while the Onkaparinga Woollen Mill[2] had emerged as the town's major employer. At the height of the Great Depression the State's Civic Record entered Lobethal as being '...a very flourishing town, having, in addition to the [woollen mill]... a cricket bat factory, and considerable support from the neighbouring farm and orchid land' (1936 cited in Young *et al.* 1983, 23). These industries provided secure employment for both Lobethal and several surrounding towns. Lobethal's insulation from the severity of the Great Depression of the 1930s is exemplified through the opening of an ambulance station (1934), police station (1935) and cinema (1936).

---

2 The Onkaparinga Woollen Company was formed in 1928. In fact this company represented the former Lobethal Tweed Company that was taken over in 1887 and renamed the South Australian Woollen Factory Company.

**Figure 9.1   German Club, Adelaide, decorated for Hitler's 50th birthday,
                 20 April 1939**

Source: National Library of Australia (nla.pic-an24460199) reproduced with permission.

While Lobethal continued to prosper economically throughout the Second World War and into the early post-war period, the stigma of Nazism was cast over the Germanic villages.[3] Fears of a Fascist fifth column within South Australia re-emerged. As during World War I, Germanic heritage marked persons as being of dubious loyalties. For many, these suspicions were fuelled by the support for the National Socialist Party and Adolf Hitler amongst South Australia's German community. As depicted in Figure 9.1, the German Club Adelaide celebrated Hitler's 50th birthday on 20 April 1939, only some five months before Britain declared war on 3 September. With the commencement of hostilities, the Commonwealth Government passed the *National Security Act 1939*, which enabled the internment of civilians, referred to as 'enemy aliens'. In South Australia, a mixture of enemy aliens and prisoners of war were held at the Loveday Internment Camp near Barmera. With the end of the war in 1945, debate turned from the interment of enemy aliens to the fear that the post war migration boom would enable hardened Nazis to enter Australia (see Sauer 1999). While certainly a legitimate concern given the number of Nazis who sought

---

3   During fieldwork, one respondent recounted how Australian soldiers from a local military base had 'raided' Lobethal in preparation for combat deployment. Verification of this story from other sources proved difficult. However, several other respondents recounted harassment during the war period on the basis of their Germanic heritage.

to escape prosecution for war crimes, this debate reified the social and political stain of Germanic heritage. Voices for restricting German migrants mobilised the notion of a '...pan-Germanic idea of racial superiority' (cited in Sauer 1999, 435) that would undermine Australia's democratic system. More bluntly, one article in the Sydney Morning Herald (20 February 1951 cited in Sauer 1999, 435) asserted that 'Germans are notorious meddlers, arrogant to a degree, and without any natural instinct for a democratic way of life'. This sentiment was not solely a product of the experience of two world wars. It echoes much earlier views of German nature and behaviour that were expressed by writers within colonial South Australia. George French Angus, son of the philanthropist George Fife Angus, wrote in his 1846 book *South Australia Illustrated* that the Germanic migrants were '...a slow, plodding class... [who] frequently exhibit considerable selfishness and ingratitude' (cited in Whitelock 2000, 71).

In the aftermath of these troubles, Lobethal began to experience a steady downturn in its economy during the 1950s. An agricultural depression during the 1970s combined with productivity declines at the Onkaparinga Woollen Mill throughout the 1980s further undermined the village. Lobethal also experienced further dilution of its strong Germanic character due to improved transport and in-migration from other areas. Lobethal's steady decline was completed by the closure of the woollen mill in 1993. Arguably, the village that had weathered the turmoil of religious schisms, xenophobia and depression while retaining its sense of heritage had reached its nadir.

Place-making – Christmas wonderland?

The closure of the woollen mill was a significant blow for the village. One respondent recalled that the closure was catastrophic as '[w]hole families worked at the mill, generations', concluding that there was not '...one family in Lobethal that didn't have some... connection with the woollen mills...[it was] devastating'. Attempts to revitalise the village's economic base had little effect. Seeding funds from the South Australian State Government were used to establish a local craft and produce market in the disused mill. Although the market continues to operate, it is viewed by the local community as a failure. Faced with the prospect of becoming yet another rural community in decline, Lobethal was fortunate to have the Germanic tradition of decorating the home at Christmas time with candles, an activity that had slowly grown into a community celebration, to draw upon. In effect, the formalisation of the Festival of Lights is a defining moment in the reinvention and reinvigoration of Lobethal.

The genesis of the present Festival of Lights can be traced back to 1947 and so has existed for some sixty years. The exact origin of the lights is hotly debated, with local history buffs crediting a direct descendant of the original Lutheran settlers with starting the practice by placing candles in the windows of her home on Christmas Eve. Alternatively, others recall a shopkeeper stringing painted lights outside his Main Street business to attract customers and to celebrate the festive season. They recounted stories of hand painting light globes and making crude strings of them to decorate building fronts. Both accounts have elements of truth. Certainly lighting one's home with candles at Christmas is a Germanic tradition that undoubtedly was transplanted to the colony with the Germanic settlers. Further, it is credible that this private ritual crossed over into the public sphere of Lobethal's main street. Regardless, what began in a small and individual way, as the rekindling of a Germanic Christmas tradition has become a nationally promoted event. Reflecting this transition, Lobethal enjoys national exposure through home-making magazines and has recently received international recognition in Christmas themed books.

So successful has the Lights Festival been, that the South Australian Government has celebrated Lobethal as a template for other ailing rural communities:

> In the early nineties when the Onkaparinga Woollen Mills closed the town's business community and residents met to discuss what could be done to restore community confidence. The town as a whole decided to establish the Lobethal Christmas Lights as the finest community display in Australia... the Lobethal Christmas Lights came about because the community was in crisis when its major employer shut down. Today the lights are a symbol of community enterprise and... pride. (http://www.communitynet.sa.gov. au/case_study.asp?Case_Study_ID=19).

Beyond the obvious boosterism, this assessment is erroneous in several respects. Residents sampled refute the united community/commerce scenario, claiming that no such organised forum occurred. This reflects Metcalfe and Bern's (1994, 665) observation that '...the past is not simply given. It is the arena for struggles over remembering and forgetting, as people try to claim the future'. This astute observation is further highlighted by the competing claims over the origins of the Festival of Lights. Interestingly, the Festival committee accepts the theory of the direct descendant over that of the local shopkeeper theory. In the context of the Festival's rural idyllic construction this is significant. Tracing the origins back to a direct descendant lends the Festival an enhanced sense of 'tradition' and rustic community values. Thus, the official history of the Festival is replete with the rhetoric of heritage that may be considered as constituting an 'invented tradition'. During the Festival of Lights, Lobethal becomes the realisation of both the rural idyll and of traditional Christmas so celebrated within Western thought. The deliberate mobilisation of specific discourses has been used to promote Lobethal literally and imaginatively as the *Valley of Praise*.

Over the Festival's seventeen days, the wider public are invited to participate in Lobethal's traditional and wholesome community. This is Lobethal's 'gift' to the wider community. As promotional materials enthuse, '...the true spirit of Christmas can be captured by... strolling through the streets of Lobethal' (Bank SA promotional brochure, 2003). This invitation encourages visitors to immerse themselves in an idyllic community setting through which they can commune with the true spirit of Christmas (Figure 9.2). This sense of communion was aptly communicated through entries in visitors' books provided at various homes. Brief comments such as 'great, it feels like Christmas now' (13 December 1998); 'you can feel the magic of Christmas' (9 December 2002) and 'you gave a real sense of Christmas spirit to me' (14 December 1998) evocatively capture the overwhelming sense of visitor appreciation. More importantly, Lobethal's Christmas tradition has become an integral aspect of numerous visitors' sense of Christmas tradition, with one visitor writing; '…it is part of our family Christmas tradition to come to Lobethal every year' (22 December 1997). Thus, the Christmas spirit and tradition enjoyed by Lobethal's residents has become a part of the Christmas tradition for the Festival's predominantly urban visitors.

Figure 9.2   Santa's Retreat: A private Lobethal home adorned extensively with Christmas lights

Source: Winchester and Rofe (2005, 268). Photograph reprinted by permission of Elsevier.

A country Christmas

Works examining Christmas as a socio-religious phenomenon are surprisingly few. Those that do exist tend to focus upon the social history of the season and/or the ritual of gift giving (see for example Connelly 1999; Miller 1993; Nissenbaum 1997; Waits 1993). Indeed, Waits (1993, 3) has gone as far as to assert that '[r]eligion has not played an important role in the emergence of the modern form of the celebration'. This is certainly not the case at Lobethal. Religion has shaped the form of the Festival considerably, drawing on Lobethal's strong Lutheran heritage. A core theme of the Festival is a re-centring of Christian values and ceremonies within Christmas celebrations, exemplified by the live Nativity play each night and the number of explicit references to the sacred origins of the Christmas celebration.

Like romanticised notions of the rural idyll, Christmas can be considered as an 'invented real tradition' (Miller 1993; Nissenbaum 1997). Arguably paradoxical, this term draws upon the intersection of traditional notions of 'CHRISTmas' with the secular, modern form of Christmas. Thus, the Christian premise of Christmas forms a traditional foundation or backdrop for the secular commercialisation of the season.

Miller (1993, 4) asserts that '[t]he Christmas ritual which we know today was the "invention" of the relatively well-to-do Victorian middle class and reflects their preoccupations'. This is vitally important as the rise of the Christmas ritual coincided with the emergence of the rural idyll. A close reading of these literatures reveals this common foundation of class longing. Nineteenth century English newspapers were redolent with articles extolling the virtues of Christmas rituals (see Connelly 1999). A central theme of these writings is Christmas as a *time* of enhanced community spirit:

> Christmas! Longed for as the season when our shining hearths, our seasonal fires, our domestic comforts and our social felicity become brightest under the Christian sun! (Fyfe 1860, cited in Connelly 1999, 11).

Here, Christmas is celebrated as a time of eroded social structures, community spirit and, most significantly, Christian communion. However, Christmas is not only a celebration in time, it was equally constructed as a celebration of a special time in *space*. Victorian depictions of Christmas located the festive season in an idyllic rural setting:

> It is hard to picture a more pleasant scene than that of an Old English Christmas. The country-side... echoed with songs of glee and merriment... In the roomy, old-fashioned houses... rich and poor discarded for a time all class distinctions and joined equally in the merry-making (*West Briton and Cornwall Advertiser*, 27 December 1900, cited in Connelly 1999, 24).

Thus, the rural is romanticised as the natural domain of traditional Christmas values. This association emerged from the same discourse demonising the urban as the landscape of social and moral decline. Just as community and kinship forms were corrupted in urban environments, so too were wholesome Christian values. Consequently, Victorian writers emphasised that:

> ...the only way... to see Christmas... surrounded by all its poetical associations, is to spend it in the country. In town it is tricked out in the new fashions – very pretty to look at, yet in nowise romantic. But *there are... out-of-the-way nooks... which, lying from off the great high roads, seem to have been forgotten by... time... These are the spots where you feel the Poetry of Christmas to its full* (*Illustrated London News*, 24 December, 1853, cited in Connelly 1999, 27) (emphases added).

Rural spaces of Christmas tradition are constructed as being apart from the urban, but equally as echoes of another, more genteel time. Hopkins (1998) refers to this as rural alterity. Rural alterity, refers to the way the '...rural is represented as being some place other than urban, as some time other than the present, as some experience other than the norm' (Hopkins 1998, 78). Despite being a physical place, during the Festival of Lights Lobethal is transformed into an alterit space in which invented traditions are placed on display. In the case of Lobethal, its rural setting and community is positioned as a natural place where the Christian emphasis of Christmas can be re-affirmed. However, caution must be exercised when unravelling the complexities inherent in alterit landscapes and times such as Lobethal's Festival of Lights. Although the overall Festival is joyous and enjoyed, there are elements of collision in the landscape between the secular and the sacred, the individual and the managed and the domestic and the foreign. These collisions are further explored in the last section of this chapter.

Fractures in the Valley of Praise

On the surface, the Christmas spirit being promoted is one of family and community. However, increased commercialisation problematises the Festival's discourse of Christmas spirit. Promotional materials extol commercial interests alongside community sentiments:

> But the experience is not just the lights. In recent years, the Lights of Lobethal Festival has encouraged a wide variety of stallholders to participate in a street market, which stretches along the main street of Lobethal. This year's Festival will be the greatest yet, with many things to see and do, and of course, there are the lights themselves (Bank SA promotional brochure, 2003).

Local residents expressed concern over the increased commercialisation of the Festival. A major fear was that over-commercialisation would dilute the Festival's community basis and meaning. Fears of the festival being reduced to a 'tacky' tourism venture were not limited to local residents alone. A number of visitors sampled also commented on the overcommercialised nature of the Festival:

> I couldn't help but think going down the main street that it [the Festival] was an asset for the commercial ventures... it just ran through my mind that they're doing well cashing in... selling ice-cream and gelato... but having turned the corner [into a residential street] and coming down here and seeing the private people involved with the spirit of Christmas... I think the real meaning of Christmas is coming through.

This excerpt not only reflects visitor concerns of commercialisation diluting the true spirit of Christmas; it also privileges the local community as the embodiment of Lobethal's Christmas spirit. This comment also strikes a chord with promotional assertions that the Christmas spirit can be 'captured' in the streets of Lobethal (Bank SA promotional brochure, 2003). Thus, Lobethal's true Christmas spirit is located in the village's residential areas as opposed to on Main Street's commercial landscape.

Overcommercialisation may well herald a decline in the popularity of the Festival in the future. Hopkins (1998) has argued that when the '...representational spaces of the symbolic countryside do not correspond to the material landscape' the idyllic rural construction can falter. Thus, fracturing of the carefully constructed nature of Lobethal as an idyllic Christmas landscape holds the potential to undermine the Festival. One resident believed that '...over a period of time that [commercial] side of it [the Festival] will die down' and that the festival would return to its community roots. Rather than being viewed with anxiety, many residents expressed a longing to return to a less formal Festival structure:

> I don't think we'll ever lose the lights, but there are some who think that in the last few years it has become too commercial with the [commercial] sponsorship and all that. They'd be quite happy if it stepped back five years, but I think the individual homes would still carry on even if the organisation itself fell down with the extra facilities and the street stalls and all that. I think it's so strong in their hearts... I can't see it ever dying out.

Indeed, these residents may well get their wish as numerous other communities in the Adelaide area are establishing Christmas lights displays. While none of these rivals the scale of the Lobethal display, they represent a more accessible option for many urban residents. Recent newspaper articles have recognised this proliferation (Clemow 2002; Merriman 2002; Quast 2003), one even boldly declaring 'Lobethal Challenged' (Lato 2003, 15). However, given the history of the Festival and the way it constitutes an important Christmas tradition for residents and so many visitors alike, the future of the Festival appears assured.

Figure 9.3   Recentralising CHRISTmas: A house displays a strong message to emphasise the Christian aspect of the festival, reinforced by the absence of lights and decorations

Source: Winchester and Rofe (2005, 270). Photograph reprinted by permission of Elsevier.

The clearest element of collision is between the secular and the sacred. Many houses adopt the familiar Santa theme with reindeer, trees and candy canes represented in lights and/or cut-outs. Others are more clearly religious, with extensive Nativity scenes, angels, stars and bells. Some individuals choose to make their religious message very clear (Figure 9.3), while others are subtler, or the messages and images are mixed. These contrasting narratives are indicative of the framing of secular notions of Christmas within the broader structure of CHRISTmas.

Residents and visitors alike held a range of perspectives regarding the centrality of Christian values and messages within the modern Christmas celebration. While some respondents placed greater emphasis on the secular form of Christmas celebrated in Lobethal, others drew inspiration from the CHRISTmas themes in many displays. Emphasising the Lutheran foundation of Lobethal's Christmas, one respondent stated:

> The *Valley of Praise* is what it's known as. People [here] just get a thrill out of just being able… to spread the message of Christmas and Christ's birth. That's what Christmas is all about and the people are trying in their own way just to let people know just how happy they are at Christ's birthday.

The spirit of Christmas celebrated at Lobethal was epitomised by the recurring Biblical quotation 'Glory to God in the highest, and on earth peace, good will toward men' (Luke 2:14). This quotation encapsulates the key themes of Victorian constructions of Christmas as a time of praise, harmony and communion. Other homes displayed blunter messages (Figure 9.3). Worthy of note is that those homes proclaiming such an overt CHRISTmas message were either devoid of lighting displays (Figure 9.3) or modestly decorated. One respondent attributed the large number of such overt CHRISTmas messages to the 'depth' of Christian commitment within Lobethal. In a similar vein, another resident drew upon an historical narrative of the village, explaining:

> …it was [originally] a Lutheran town and those Lutheran pioneer families are still here in large numbers, so it's a fairly strong Lutheran community. And there are all the other churches that are quite strong, so I think it would be true to say that there is a very Christian underlying… particularly in the older people. I think you'll find that they tend to be offended if you drift away from that Christian meaning of Christmas.

In this respondent's mind, the village's Lutheran origins centralises Christian traditions and values within the community's celebration of Christmas.

Conclusions

This chapter has examined the intersection of the German Lutheran heritage of the village of Lobethal in the Adelaide Hills with the romanticised rural idyll. The rural idyll has traditionally been conceptualised as a place of tranquillity, harmony, and community, the antithesis of the urban industrial sprawl which was conducive to crime, overcrowding and poverty. This myth has been problematised in recent years, with an increasing acknowledgement of the tensions and complexities of rural life.

In Lobethal, the rural idyll was generated by the perception of the village as a refuge haven, where the refugees themselves were deemed to be pious and hardworking, and where their environment in the Adelaide Hills was perceived to be cooler, fresher and purer than that of the City of Adelaide. In this chapter, we have contrasted this idyllic myth with some of the realities engendered by the village's Germanic heritage. In particular, the settlement era may be seen as a time when refugees were exploited and made indebted to enhance the avaricious claims of landowners. The establishment of the village itself also resulted from religious differences from within the Germanic Lutheran community. The stark contrasts of refugee haven on the one hand, or zone of exploitation on the other, are overstated here, but serve to identify the multiple views of a complex reality of Lobethal's establishment.

Throughout the first half of the twentieth century, the eras of the two World Wars brought the myth of the pious hard-working German settlers into stark contrast with the stereotype of Germanic aggressors responsible for countless atrocities. As a consequence of these wars, a number of 'enemy aliens' were held in internment camps and the Germanic name of Lobethal was expunged from the map for nearly twenty years. It is notable at this time that the romanticised rural myth appears to have been in abeyance. The Germanic Lutheran community became demonised as a consequence of events occurring half a world away, and so the place could no longer be seen as one of tranquillity and purity. Indeed, for a while this Germanic place ceased to exist and Lobethal took on the guise of Tweedvale.

The rural idyll again combines with the Germanic Lutheran heritage in the post-war years as a place-making strategy. The Germanic tradition of lighting up homes with coloured candles forms the basis of Lobethal's Christmas Festival of Lights. The place-making construction is built on a flimsy detail of Germanic heritage, whereas by the late twentieth century, the stereotypes of both the pious industrious settler, and the enemy aggressor have become much more muted. The development of this tradition into a commercially-sponsored event draws on romanticised notions of Christmas as a time of community spirit best experienced in the countryside. During the seventeen days of the Festival, Lobethal is physically transformed into a symbolic manifestation of a highly nostalgic rural idyll. However, the exact origin of the Festival is disputed and its commercialisation is contested. The landscape reflects the range of views from those whose religious beliefs and heritage are deep-seated to those who are more comfortable with secular versions of Christmas. Although the Festival of Lights may be seen as an invented tradition, it has been effective as a place-making strategy, which generates large numbers of visitors and national media coverage.

This chapter has focussed on two main constructions, those of the Germanic Lutheran heritage and of the rural idyll, both of which are used in complex ways to create the meanings of the place of Lobethal. The rural idyll has been utilised throughout the history of Lobethal, except at times when the community itself was demonised and stereotyped. The Germanic Lutheran heritage of Lobethal is seen as contested and partial at every stage from the establishment of the religious haven through its role as seditious hideout to its current symbolic place as a Christmas wonderland through the Festival of Lights.

References

Australian Bureau of Statistics (2002) *Changing of German Place Names in Australia*, Year Book Australia, Culture and Recreation, 1301.0.

Bessière, J. (1998) 'Local Development and Heritage: Traditional Food and Cuisine as Tourist Attractions in Rural Areas', *Sociologia Ruralis* 38:1, 21–34.

Brauer, A. (1985) *Under the Southern Cross: History of the Evangelical Lutheran Church of Australia* (Adelaide: Lutheran Publishing House).

Bunce, M. (1994) *The Countryside Ideal: Anglo-American Images of Landscape* (London: Routledge).

Clemow, M. (2002) 'Len Lights the Way for Santa', *Sunday Mail* South Australia (22nd December), 64.

Cloke, P. and Little J. (1997) (eds) *Contested Countryside Cultures: Otherness, Marginalisation and Rurality* (Routledge, London).

Connelly, M. (1999) *Christmas: a Social History.* (London: I.B. Tauris Publishers).

Dunn, K.M., McGuirk, P.M. and Winchester, H.P.M. (1995) 'Place Making: the Social Construction of Newcastle', *Australian Geographical Studies* 33:2, 149–167.

Dutton, F. (1846) *South Australia and its Mines, with an Historical Sketch of the Colony* (London: T. and F. Boone).

Ekman, A. (1999) 'The Revival of Cultural Celebrations in Regional Sweden. Aspects of Tradition and Transition', *Sociologia Ruralis* 39:3, 280–293.

Foster, R., Hosking, R. and Nettelbeck, A. (2001) *Fatal Collisions: The South Australian Frontier and the Violence of Memory* (Adelaide: Wakefield Press).

Frankenberg, R. (1975) *Communities in Britain: Social Life in Town and Country* (Harmondsworth: Penguin).

Gerber, D.A. (1984) 'The Pathos of Exile: Old Lutheran Refugees in the United States and South Australia', *Comparative Studies in Society and History* 26:3, 498–522.

Halewood, C. and Hannam, K. (2001) 'Viking Heritage Tourism: Authenticity and Commodification', *Annals of Tourism Research* 28:3, 565–580.

Hansen, K. (1999) 'Emerging Ethnification in Marginal Areas of Sweden', *Sociologia Ruralis* 39:3, 294–310.

Hopkins, J. (1998) 'Signs of the Post-rural: Marketing Myths of a Symbolic Countryside', *Geografiska Annaler* 80:B, 65–81.

Jackson, P. (1989) *Maps of Meaning: An Introduction to Cultural Geography* (London: Unwin Hyman).

Lato, D. (2003) 'Lobethal Challenged', *The Advertiser* (23 December), 15.

Lewis, P.F. (1979) 'Axioms for Reading the Landscape: Some Guides on the American Scene', in Meinig, D.W. (ed.) *The Interpretation of Ordinary Landscapes: Geographical Essays* (New York: Oxford University Press), 11–33.

Little, J. and Austin, P. (1996) 'Women and the Rural Idyll', *Journal of Rural Studies* 12:2, 101–111.

Matthews, H., Taylor, M., Sherwood, K., Tucker, F. and Limb, M. (2000) 'Growing-up in the Countryside: Children and the Rural Idyll', *Journal of Rural Studies*, 16, 141–153.

Merriman, J. (2002) 'Streets Ahead of the Rest', *Sunday Mail* South Australia (15 December), 25.

Metcalfe, A.W. and Bern, J. (1994) 'Stories of Crisis: Restructuring Australian Industry and Renewing the Past', *International Journal of Urban and Regional Research* 18, 658–672.

Miller, D. (ed.) (1993) *Unwrapping Christmas* (Oxford: Clarendon Press).

Mingay, G.E. (ed.) (1989) *The Unquiet Countryside* (London: Routledge).

Newby, H. (1987) *Country Life: A Social History of Rural England* (New Jersey: Barnes and Noble Books).

Nissenbaum, S. (1997) *The Battle for Christmas* (New York: Alfred A. Knopf).

Paddison, R. (1993) 'City Marketing, Image Reconstruction and Urban Regeneration', *Urban Studies* 30:2, 339–350.

Panelli, R., Stolte, O. and Bedford, R. (2003) 'The Reinvention of Tirau: Landscape as a Record of Changing Economy and Culture', *Sociologia Ruralis* 43:4, 379–400.

Quast, J. (2003) 'Traders Plan to Light up Port Adelaide', *Messenger – Portside* (26 March), 9.

Rofe, M.W. (2004) 'From "Problem City" to "Promise City": Gentrification and the Revitalisation of Newcastle', *Australian Geographical Studies* 42:2, 193–206.

Sauer, A.E. (1999) 'Model Workers or Hardened Nazis? The Australian Debate about Admitting German Migrants, 1950–1952', *Australian Journal of Politics and History* 45:3, 422–437.

Short, J. (1991) *Imagined Country: Society, Culture and Environment* (London: Routledge).

Sibley, D. (1995) *Geographies of Exclusion: Society and Difference in the West* (London: Routledge).

Smith, S.J. (1993) 'Bounding the Borders: Claiming Space and Making Place in Rural Scotland', *Transactions of the Institute of British Geographers* 18, 291–308.

Stacey, M. (1969) 'The Myth of Community Studies', *British Journal of Sociology* 20, 134–146.

The Register (1916) 'In the Assembly', 3 August, 2.

The Register (1916) 'Removing German Names: Difficult and Delicate Task', 5 August, 13.

Tonts, M. and Grieve, S. (2002) 'Commodification and Creative Destruction in the Australian Rural Landscape: The Case of Bridgetown, Western Australia', *Australian Geographical Studies* 40:1, 58–70.

Waits, W.B. (1993) *The Modern Christmas in America: A Cultural History* (New York: New York University Press).

Whitelock, D. (2000) *Adelaide: From Colony to Jubilee, A Sense of Difference* 3rd Edition. (Adelaide: Arcadia Press).

Winchester, H.P.M and Rofe, M.W. (2005) 'Christmas in the "Valley of Praise": Intersections of the Rural Idyll, Heritage and Community in Lobethal, South Australia, *Journal of Rural Studies*, 21, 265–279.

Woodruff, P. (1984) *Two Million South Australians* (Adelaide: Peacock Publications).

Young, G., Aeuckens, A., Green, A. and Nikias, S. (1983) *Lobethal (Valley of Praise)* (Adelaide: South Australian Centre for Settlement Studies).

Chapter 10

Perth's Commonwealth Games Heritage: Whose Value at What Price?

Catherine Kennewell and Brian J. Shaw

Prologue

In 2005, on the first day of the 'traditional' Boxing Day cricket Test Match at the Melbourne Cricket Ground (MCG), the Federal Treasurer Peter Costello announced the inclusion of that stadium on Australia's National Heritage List. Championed as the 'home of Australian sport' the stadium has a time-honoured association with the Australian Rules football code, both interstate and international cricket, and was the venue for the 1956 Olympic Games. Although the theatrical nature of the public announcement might well be regarded as political opportunism, official recognition of the MCG bears testimony to the pivotal role played by sport within contemporary Australian society. While the egalitarian image of the 'bronzed Aussie' competitor has been acknowledged to be somewhat of a mythical figure, Australians from all walks of life remain most enthusiastic consumers of sporting spectacle, whether live or transmitted (McKay 1991; Real 1998). Indeed, it is the tradition of sporting spectacle that is being recognised through the listing of the MCG, with the stadium itself lacking architectural authenticity having been constantly rebuilt and refurbished over a period of 150 years, most recently as the major venue for the 2006 Commonwealth Games.

Such recognition bears directly upon the theme of this chapter, which explores the value and cost of securing and retaining sporting heritage in a rapidly changing recreation and leisure environment, placed within a rapidly changing urban and socio-economic environment in which sport has succumbed to ever increasing commercialisation and commodification. Giulianotti (2002, 29) has termed this 'hypercommodification' whereby extraordinary and different volumes of capital have entered the market and affected the relationship between sport and its followers. Within today's commercialised, corporatised game, tradition and history count for little unless they add value to contemporary marketing and promotion strategies. In some cases where tradition has been absent it has been invented. Thus the advent of the aforementioned MCG Boxing Day cricket Test Match only dates from the 1980s when playing schedules were redesigned to meet the needs of Kerry Packer's media empire. Alternatively, the downside of this process can be seen in the demise of the interstate Sheffield Shield competition, inaugurated in 1892 when Lord Sheffield donated a trophy 'for the betterment of Australian cricket' (Shaw 1984, 152). The abandonment of this established icon occurred abruptly in mid-season 1999–2000, after the Australian Cricket Board (ACB) sold out the naming rights to National Foods Limited who substituted their own 'Pura Cup' competition, named after a brand of milk.

Therefore, while sporting tradition has been publicly honoured through the high profile listing of the MCG, the imminent national recognition of other stadia or sporting heritage around Australia remains somewhat improbable. Some difficulties involved in the elevation of such places are examined below through two case studies of heritage sites dating from the VIIth British Empire and Commonwealth Games held in Perth, Western Australia, from 22 November to 1 December 1962. Perth's Games legacy represents much more than 'just' sporting heritage. The Games were instrumental in exposing the rather parochial capital of Australia's hitherto 'Cinderella State' to the much wider world. At the time the organisers claimed that the Games generated the greatest sustained level of overseas publicity since the Gold Rush days of the 1890s (Gregory 2003, 92). Yet, despite the historical importance of the Games in raising Perth's profile, the future of many of its surviving structures is decidedly uncertain due to a concatenation of vested interests and urban planning initiatives. Bound up with questions of heritage conservation in this context are issues that relate to the upgrading and proposed relocation of sporting stadia, as well as the reconstitution of the established built environment and the socio-economic character of the (now) inner suburban areas in which the structures are located.

The VIIth British Empire & Commonwealth Games

The earliest recorded games between Empire athletes coincided with celebrations for the Coronation of King George V at the Crystal Palace Grounds in London, 1911. It was known as the 'Festival of the Empire', a series of entertainments and exhibitions relating to the progress and development of the British Empire featuring an inter-Empire Sports meeting. The first British Empire Games *per se* were held in Hamilton, Canada in 1930, and thereafter in London (1934), Sydney (1938) and Auckland (1950), before being renamed the British Empire & Commonwealth Games, which took place in Vancouver (1954) and Cardiff (1958) (see Australian Commonwealth Games Association 2004). Perth, with a population of 420,000 in 1961, was encouraged to bid for the 1962 VIIth meeting although the city had no sports facilities of an international standard and had never staged a major international sporting event. The application itself was somewhat controversial, with Adelaide initially gaining support from the Australian British Empire and Commonwealth Games Association before Perth lodged a successful protest. In the event Perth provided better facilities than any previous meeting, organised a wide promotional campaign, and consequently attracted the largest audience of any Empire & Commonwealth Games to that date, for an overall outlay of around £4million (Edmonds 1962; Edmonds and Willmott 1962).

No less than eleven separate venues, all lying within 10kms from the city centre, were chosen as sites for the various activities (see Figure 10.1). Some, such as the Dalkeith Bowling Club and South Perth Civic Centre (weightlifting), were already existing facilities and others like Kings Park (road cycling) and the Canning River (rowing) took advantage of Perth's natural setting. Four new structures were custom built for the occasion; the cycling velodrome at Lake Monger was opened prior to the Games in 1959; the aquatic centre was located at Beatty Park but initially planned for Kings Park; the main athletics stadium was at Perry Lakes; and the Games Village in City Beach. Of these newly constructed facilities the initial proposal for the aquatic centre was most controversial, since the Council endeavoured to excise protected land in Kings Park in order to accommodate the facility on some 17.75 acres (7ha) within the 1,000acre park. In the face of public opposition this proposal was eventually defeated in parliament and the organisers were obliged to find an alternative site for the pool as a matter of some urgency. After further deliberations Beatty Park, north of the Perth CBD, was selected, despite some local objections, and work began on clearing the site in May 1961 for a structure which covered just 4.75 acres at a cost of £640,000 (Edmonds and Willmott 1962; Gregory 2003, 80–85).

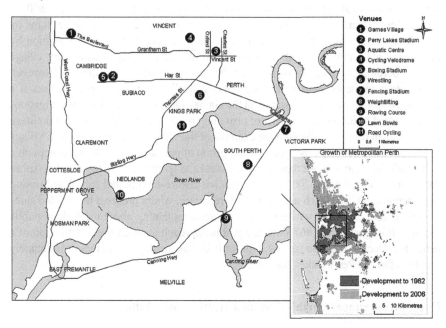

Figure 10.1   Perth British Empire & Commonwealth Games venues, 1962

Perry Lakes stadium, the main venue for track and field events at the Games was sited in the newly developing suburb of Floreat Park, some 6kms from the city centre, on the former Lime Kilns Estate which had been purchased by Perth City Council from local landowner Joseph Perry in 1917. The estate was incorporated into the City's Endowment Lands, which had been originally designed to provide income for municipal works through activities such as firewood collection, stone quarrying and grazing. Subsequently, the City of Perth Endowment Lands Act 1920 empowered the City to develop and sell this land in its trust, and the first land sales took place in City Beach in 1929, and Floreat Park in 1934 (Town of Cambridge

1997). The stadium, completed in 1962 at a cost of some £700,000, was located in the southeast corner of the former estate and the surrounding parklands were used for practice tracks, various associated playing fields and, during the period of the Games, public car parks. Designed to be state-of-the-art, the concrete frame stadium boasted an innovative scoreboard capable of showing both numbers and letters, a communications centre to accommodate two hundred journalists and technical facilities for both radio and the fledgling television coverage (Gregory 2003, 79). The stadium was designed to have a capacity of 30,000, fully seated, but additional temporary stands were added during the Games to extend this beyond 35,000. However, attention to spectator comfort reflected the standards of the times, seating was on wooden benches and no protection was provided from the elements, apart from 3,000 places located under a cantilever-roofed pavilion (Edmonds 1962).

Accommodation in close proximity to the stadium was deemed necessary to house over one thousand competitors and officials, and, in May 1959, Perth City Council granted 76acres (30ha) of its undeveloped endowment land for the Games Village site at City Beach. The site gave glimpses of the Indian Ocean, was some 4kms by road from the stadium site and 10kms from the city centre. The first suggestion by the council in conjunction with the government was to construct a temporary village of transportable housing, which could be removed after the Games. However, surrounding residents were concerned about a decline in their property values if a temporary village were to be located on the Games site and thus detailed investigations were initiated to determine the cost of providing roads, water, electricity and drainage for permanent housing following the example of the 1956 Heidelberg Olympic Village in Melbourne. The plan was to construct a group of permanent private residences on a new subdivision, and to use them for a few weeks as temporary Games housing before they were sold (Stickells 2002). This initiative also re-invigorated suburban housing development in City Beach and nearby Floreat Park, where progress had stalled since the initial land sales and where residents were complaining about the lack of public facilities (Gregory 2003, 86). Ultimately, this approach avoided the construction of a dedicated specific-purpose facility with limited future utility, a fate that unfortunately was to befall the Perry Lakes stadium.

Unwanted heritage? Perth's Commonwealth Games Village

Proposals for the layout of the Village were thrown open to public submission, whereas house design was to be decided through an architectural competition. The successful Village layout placed low-density residential areas on the sides of a flat amphitheatre of gently folding ground, surrounding a central park area of five hectares. This was very much in tune with the garden city planning themes already adopted in Floreat Park and City Beach (Freestone 1989). Forty architects submitted a total of 166 designs to the architectural competition and the winning entries reflected the nature of contemporary modernist architecture (see Figures 10.2 and 10.3). Floating roofs, wide skylights, open courtyards, and expanses of window were common architectural attributes. Some houses included the latest trends, such as an informal family room for TV viewing as well as a formal lounge for receiving

**Figure 10.2   Perspective sketch of Silver, Fairbrother and Associates B1 house
First Place Getter**

Reproduced with permission of Department of Housing Works, Western Australia.

**Figure 10.3   Perspective sketch of Cameron, Chisholm and Nicol B1 house
Second Place Getter**

Reproduced from microfiche collection with permission of Cameron, Chisholm and Nicol.

visitors, and most possessed either a carport or a garage (*Daily News* 14 June 1961).
A variety of timber, brick and stone materials were used, and houses were painted
in different colours so there was no look of bland uniformity throughout the Village
(Perth Jaycee Publications 1962).

At the conclusion of the Games, temporary buildings such as the gateway,
administration block, dining rooms, kitchen and recreation hall were removed
from the central parkland, while perimeter fencing was placed around the Village

Figure 10.4   A typical example of a surviving Games Village house

Photograph by the author.

to protect the state's investment during the renovation period. Finally, six months after the last athletes had left, the houses were occupied by private homeowners who paid prices ranging from £4,900 to £6,785, typical of better properties at that time (Edmonds and Willmott 1962). As more houses were added to the surrounding suburb the Village blended in with its immediate surrounds, and homes were altered, renovated and added to, to suit changing tastes and trends. Over time, degradation of the original building materials became noticeable (see Figure 10.4), and the declining availability of empty blocks within the increasingly prestigious coastal suburb led to the demolition of some Village houses in favour of newer style homes (see Figure 10.5). Nevertheless, as a demonstration of modernism, the significance of the site remained. Writing in 1982, academic Duncan Richards (2003) argued that:

> The Games Village …provided an opportunity 'to show what modern architecture could do' in the production of a total environment rather than providing individual 'gems' within an established urban or suburban matrix. Time has dimmed the lustre of many of these 'gems' but it is possible to suggest that the high public reputation of the local architectural profession during the 1950s and 1960s was largely due to the impact of these effectively designed small houses.

In July 2003 the entire Games Village precinct in City Beach was referred to the Heritage Council of Western Australia as worthy of inclusion on the State Register of Heritage Places, although, by that time, just 74 of the original 150 Games homes had survived in their original, or near original state. Community concern arose when 'Villagers' received letters from the Heritage Council notifying them the precinct was

Figure 10.5   Recently upgraded housing in the village

Photograph by the author.

being assessed, and inviting opinions and information on the cultural significance of their houses (*The Post* 26 July 2003). As before, the potential loss of now vastly increased property values troubled owners who would be unable to demolish or significantly alter their houses if registration were to take place. In an immediate reaction to this concern many owners of original houses applied to the local council (Town of Cambridge) for demolition licences (*The Post* 23 August 2003) (see Figure 10.6). Thirty such applications were lodged in the six months from July to December 2003, and while only half of the licences taken out at this time were actually used to demolish properties, the impact of the potential listing was to reduce pre-emptively the surviving heritage value of the site. Additional demolitions, albeit at a reduced rate, have occurred since that time, the situation in early 2006, with just 53 original homes remaining, is shown in Figure 10.7.

In addition to proposed and actual demolitions, Villagers engaged in organised protest, which took the form of classical NIMBY (Not-In-My-Backyard) activism on the part of residents living in one of the city's wealthiest and most socio-economically homogenous suburbs.[1] The householders petitioned parliament to

---

1   In the 2001 census 58 per cent of City Beach households had incomes over $52,000 per annum compared with the Perth Metropolitan Area average of 34 per cent; 66 per cent of the labour force was classified as Professional/Managerial compared with 39 per cent average for the Metro area; 36 per cent had university qualifications compared with 14 per cent Metro area average (Australian Bureau of Statistics, *CDATA* 2001).

Actual Demolitions versus Demolition Licences

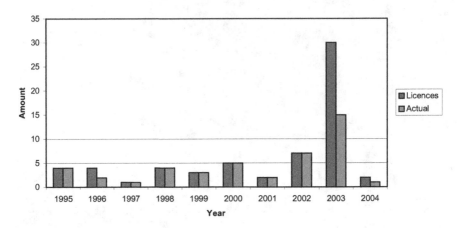

Figure 10.6   Proposed and actual demolitions, Commonwealth Games Village

Source: Town of Cambridge, reproduced with permission.

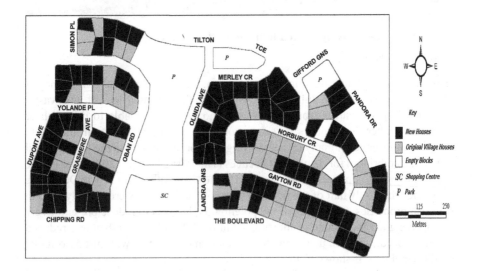

Figure 10.7   Commonwealth Games surviving properties

Source: Town of Cambridge Town Planning Map. City Beach from website. Lot numbers and data from Planning Officer, Town of Cambridge.

object to the proposed listing (*The Post* 13 September 2003); held a community rally at the local Beecroft Park, during which residents expressed their views on the heritage value of the area (*The Post* 10 January 2004); helped put together their own heritage submission to the Heritage Council (*The Post* 24 January 2004);

and invited the Heritage Minister to personally intervene in their case (*The Post* 10 January 2004 and 24 January 2004). In the midst of such pressure, local press reports appeared to indicate that the heritage precinct was to be narrowed down to twenty, and then seventeen houses (*The Post* 29 November 2003 and 13 December 2003). The final seventeen houses were to be found in the three surviving groups of four or more adjoining lots left without modification or replacement. Finally, on 13 February 2004, the Heritage Council rejected the proposal on the grounds that 'cohesion within the three nominated precincts was not strong and the integrity of the site had been undermined' (Chairman, Heritage Council pers. comm.). The news was reported in *The West Australian* 14 February 2004, and, to mark the occasion, some residents held a celebration in the local park.

Redundant heritage? Perry Lakes Stadium

On 22 November 1962, Perry Lakes stadium was filled to capacity for the opening ceremony of the VIIth Empire and Commonwealth Games. The following day's headlines in *The West Australian* (23 November 1962) read '50,000 Cheer Games Start' and, in the now somewhat dated idiom of the times, described the scene as 'a Gay Opening'. Official attendance figures record that 34,496 tickets were sold for the opening ceremony bringing receipts of £54,300 (Edmonds and Willmott 1962, 194). Unfortunately the opening days of the Games coincided with unseasonably hot weather and further within the newspaper it was recorded that 'widespread casualties were cases of heat exhaustion and collapse caused by sitting and standing in the hot sun without shade' (*The West Australian* 23 November 1962, 28). The heat continued and the first day of athletics competition at the stadium, on Saturday 24 November 1962, saw mass absenteeism on the part of ticket holders. *The Sunday Times* (25 November 1962) headline read '10,000 Seats Unsold: Heat Hits Games' and *The West Australian* (26 November 1962) informed its readers that the 'stadium was half empty'. Inclement weather of a different kind marred the events a few days later and *The West Australian* (30 November 1962, 2) reported that 'lightly clad spectators ran for cover when the rains came ...the supply of pass-outs ran out ...people rushed out of the gates for raincoats left in cars'.

Notwithstanding the success of the Games, and the record crowds established for an event of its kind, these early incidents fuelled a debate over future functions and funding that began immediately after the event was concluded. One day after 34,288 people had attended the closing ceremony *The Sunday Times* (2 December 1962, 2) assured the public that the 'Stadium Won't Change' and *The West Australian* (3 December 1962, 34) followed with 'Views Differ on Stadium's Future Tenant'. *The West Australian* reported that members of the Perth City Council saw the future of the stadium as providing a 'citadel' for athletics within the state, while *The Sunday Times* outlined a more eclectic vision, viewing the stadium as a site for contests, carnivals and receptions involving not only athletics but other sporting codes such as basketball, hockey, rugby and soccer. Rather presciently, *The West Australian* (ibid.) ventured the opinion that:

> Consensus of opinion is that a tug-of-war will still start over the use of the stadium and if it is not made a football headquarters it will become a white elephant ...football is seen as the one assured source of revenue that would maintain the stadium.

However, since their formation in the 1890s, the Western Australian Football League (WAFL) clubs had developed their own grounds, mostly within inner-city locations close to traditional working class supporters and rail links. While some movement did take place after World War II when the East Fremantle team relocated to that suburb and Perth constructed a new stadium at Lathlain, again in the inner city, changes in the geography of football stadia were remarkably small in a city that grew from some 100,000 population in 1900 to 1 million by the early 1980s (Jones 2002). For WAFL supporters each stadium inspired a sense of topophilia, being sentiments arising from 'home-like, religious, scenic and historic connotations' (Bale 1993, 93) and any proposal to relocate to the western suburbs would most likely have incurred the wrath of fans. The accepted venue for finals and interstate football, Subiaco Oval, possessed a rail link and greater crowd capacity than the stadium at Perry Lakes, without the disadvantage of an athletics track that had the effect of distancing fans from the immediacy of sporting action. Indeed, as early as 1978 both rugby and soccer bodies were pleading for a purpose-built rectangular stadium to meet the requirements of their own specific codes (Investigating Committee 1978).

If its failure to become part of an evolving football tradition stalled the emergence of topophilia with respect to the Perry Lakes stadium, there would be no such hindrance to the development of its counterpart, topophobia, created by the potential negative externalities associated with the gathering of large crowds (Bale 1993, 94). On match days stadiums generate a nuisance field in which noise, traffic congestion, parking problems, and possible hooliganism/vandalism are experienced by nearby residents. While the familial nature of Australian Rules football may have substantially lessened the threat of the latter, the lack of a rail link coupled with the rapid growth of an increasingly car-dependent city, particularly at weekends, would have potentially flooded nearby streets with cars and supporters. Such regular disruption would have been at odds with the growing residential exclusivity of adjacent Floreat Park, where the garden city principles embodied in the Endowment Lands project stressed the amenity value of its undeveloped open space as a buffer between its low-density residential suburbs. Critically the stadium had the potential to impact upon residential values, a situation that would have pushed Floreat's articulate middle-class suburbanites into the type of NIMBY activism seen in the case of the Games Village.

In the absence of a major tenant, the utilisation of the Perry Lakes stadium, rather than developing as a tug-of-war situation, became more of an all-comers smorgasbord. Over time the stadium and its pavilion premises have hosted a number of sports including A-League athletics; basketball, rugby union and soccer; together with various educational sporting associations and the WA Institute of Sport. It has been the venue for occasional one-off events such as soccer matches against visiting English teams and, perhaps most famously, the Rolling Stones 'Voodoo Lounge' concert on 8 April 1995. However, none of these tenants had the capacity to fill the stadium and swell gate receipts on a regular basis, conditions that were essential to finance the stadium's upkeep and possible future refurbishment. Finally, the stadium's fate may have been sealed in 1987 with the decision to locate the state's first national Australian Football League (AFL) team at the sport's headquarters at Subiaco Oval. Subsequent ground improvements and the introduction of floodlighting at Subiaco

Oval, following similar developments at the WA Cricket Association (WACA) ground in East Perth, further widened the growing gap in spectator comforts and the ability to stage sporting spectacle that had emerged between those venues and the increasingly obsolescent Perry Lakes stadium (see Figures 10.8 and 10.9).

Figure 10.8   A view of Perry Lakes Stadium 2006

Photograph by the author.

Following the State Liberal Government's dissolution of the Perth City Council in 1993, freehold ownership of the stadium was passed to the Town of Cambridge, one of three newly constituted municipalities.[2] Essentially this move freed the city centre from the control of suburban electors but raised the spectre of creating asset-poor municipalities in the now separated suburbs that might struggle to maintain community facilities (Gregory 2003, 331). But just as a prized heirloom can be devalued outside the family lineage, it might be argued that breaking the nexus between the Perth City Council and the VIIth Empire & Commonwealth Games venue undermined the stadium's historical value in the eyes of the wider community. Cambridge, as an amalgam of the four suburbs of City Beach, Floreat, Leederville and Wembley, certainly possessed a robust rate base but carried within its boundaries

---

2   The restructuring created a City of Perth, essentially the central city area with some 5,000 residents, and three new municipalities in the Town of Cambridge, Town of Vincent and Town of Victoria Park, in effect reversing the 'Greater Perth' amalgamations which had taken place in the early C20th (see Jones 1979; Stannage 1979, 293ff).

responsibility for 5.75sq kms of parks, gardens and reserves within a total area of 22sq kms. Faced with the upkeep of its rotting wooden seats and outdated facilities, Cambridge was more inclined to view the heritage value of the stadium as a liability rather than an asset.

Figure 10.9   Spectator facilities Perry Lakes Stadium 2006

Photograph by the author.

In compliance with the requirements of the *Heritage of Western Australia Act 1990* (Section 45), the Town of Cambridge was obliged to compile a Municipal Heritage Inventory (MHI) of places with cultural heritage significance. Within this list Perry Lakes is described as 'a place of identified Aboriginal heritage significance' and Perry Lakes stadium possessed high levels of integrity and authenticity. According to the MHI (Town of Cambridge 1997) the stadium should be valued for its:

> *Historical* significance for associations with a major international sporting event held in Perth in 1962. *Social* significance for the people of Perth and in particular for those associated with athletics. Social significance for the many sporting carnivals held there over the years. *Aesthetic* significance as an example of a major sporting venue ... following the architectural style of the times.

Ultimately, however, within the provisions of the Act, decisions regarding inclusion, management and protection within the municipal inventory are vested in the local authority. In August 1998, within a year of receiving the MHI, Cambridge Council determined that Perry Lakes Stadium would no longer be used as a sporting or recreation venue, a decision precipitated by ongoing maintenance outlays and anticipated costs of refurbishment that were deemed to be prohibitive. The 15.5ha site would be reclassified from 'Parks and Recreation Reservation' to 'Urban' in a move which would entail the demolition of the existing facilities prior to land development which would potentially be most favourable to the Town's rate base (Town of Cambridge 2005).

Initial public reaction to the news of the stadium's demise was muted, to say the least. Most press reports implied that the decision was inevitable and discussion revolved around the nature and feasibility of the Town's plans for the site, rather than the loss of valuable heritage. As development plans were revealed, many public concerns arose over the Council's choice of development partner; the character and density of residential development; the financial burden of development and construction and its potential impact upon taxpayers; corruption and possible conflicts of interest among Council Members; the competency of the Town of Cambridge to handle the project; continual delays in decision making and a host of other issues, still ongoing at the time of writing (*The Post* various dates). However, the retention of the stadium for its innate heritage value was not a subject for much consideration. One ratepayer queried plans to demolish a stadium itemised on the Council's own heritage list (Town of Cambridge Minutes 22 February 2005), but this was a lone voice unrelated to the dominant community concerns. To paraphrase Peter Howard (2003, 89) it appeared that, for longstanding Perth citizens, heritage was the memory of activity surrounding the Games event and not the obsolete stadium.

Games heritage: whose value? what price?

These explorations, one of frustrated heritage management and another of apparent unconcern, have highlighted two of Perth's (thus far) better-preserved legacies of the 1962 VIIth British Empire & Commonwealth Games. Certainly other structures are still in existence, such as the Beatty Park aquatic centre and the cycling velodrome at Lake Monger, but survival in these cases has been tinged with greater levels of pragmatism. The aquatic centre has been substantially remodelled and bears little resemblance to the Games original, while the velodrome, now separated from Lake Monger by an urban freeway, has shed its cycling track and hosts a local soccer team. The fundamental issues which arise in this regard illustrate the perpetual tension which exists between authenticity and expediency; that is between the values placed upon retaining elements of the past and the real or imputed costs associated with their conservation. Implicit here, as in all cases of heritage identification and conservation, are the fundamental questions of who decides, and who pays? Herein are found the inevitable qualities of self-interest, divisiveness and contestation that define the heritage industry.

In 1962, for a few days in November and December, the Games placed Perth in the international spotlight. In this sense the whole community was somehow legitimised by the event, which conveyed the essence of Perth and Western Australia to the wider world. However, hectic progress in the following years served to discount the memories of small town Perth as a succession of mineral booms, entrepreneurial excesses and growing self-confidence fuelled economic development and rapid population growth which has now more than trebled the 1962 population (Gregory 2003). Today, while older members of the community might reflect on the Games with some nostalgia, many younger people are more likely to cite the 1986–87 America's Cup Defence as a pivotal sporting moment in the development of the city (Syme et al. 1989). Indeed, it may be argued that the heritage value of the latter has far exceeded that of the Games, with the City of Fremantle having been particularly successful in capitalising on that event to forge a new and distinct identity. America's Cup notwithstanding, Perth's inability to attract further hallmark events has reduced the opportunity to showcase and possibly renovate Perry Lakes stadium, in the manner of the national heritage listed MCG or, even more appealingly, the Los Angeles Coliseum which has now hosted two Olympiads,

If fading memories of the stadium no longer evoke a sense of belonging and identity for Perth residents, the Games Village was likely to have an even more tenuous emotional attachment to the broader community. The majority of Perth residents would have little knowledge of the Village, situated as it is within elite City Beach, a suburb that has steadfastly maintained its exclusiveness over time. Moreover, affluent and well-qualified Village residents themselves had little need for the cultural capital that such association might bring. However, any objective assessment of heritage value for Villagers quickly disappeared with the growing fears of much reduced resale values or of restrictive development guidelines on their privately owned properties. In this regard a survey of City Beach residents, conducted in 2004 after the rejection of the heritage proposal, revealed fine spatial distinctions in the moulding of community attitudes. Opinions were solicited on both the perceived heritage value of the Village and on residents' personal experiences of the nomination process.[3] The survey findings revealed that, in terms of personal involvement in protest actions, 90 per cent of 'Inner Village' residents participated in such action while 89 per cent of 'Outer Village' residents did not. Furthermore, while 74 per cent of all respondents thought the Village was 'not worthy of heritage status', all residents of heritage nominated properties concurred with this view and Inner Village respondents were more likely to discount the heritage value of their properties than were Outer Village respondents. Overall, the nuance of opinion changed over very small distances, and at the scale of adjoining streets.

Similar differences in responses were found when residents were asked if heritage listings should be applied to suburban properties. Sixty-five per cent did not agree with such listings, but Outer Village residents were less likely to oppose

---

3   The survey was composed of 133 mail-back questionnaires sent to three groups of households; firstly those directly affected or 'nominated' residents; secondly other 'inner' Village residents living on the original 150 blocks; and thirdly 'outer' Village residents beyond the original Games Village. Fifty-eight households responded, a response rate of 43.6 per cent. For the purposes of analysis 'Inner Villagers' including nominated and original block households were compared with 'Outer Villagers' (McDonald, C. (2004) *Perth Heritage and the 1962 Commonwealth Games Village*, unpublished Honours Thesis, School of Earth and Geographical Sciences, The University of Western Australia).

such moves. Asked whether they would participate in protests again at some time in the future, most respondents did not like the implication that the Village could be re-listed, but while most Inner Village respondents stated that they would be involved again, 60 per cent of Outer Village respondents were disinclined to do so. When asked what should be done to commemorate the Games, in the absence of heritage listing, both Outer and Inner Village respondents favoured a restored model home, or preservation of the Perry Lakes stadium, while all respondents within heritage nominated properties wanted just an information display or indeed nothing at all. Overall the survey tended to reinforce the notion that heritage values are subjective and conditional. Spatially this translated into a 'core' area (Inner Village), which was negatively disposed to any commemoration, and a more disassociated periphery (Outer Village).

Postscript

For some time, WA state governments of diverse political persuasions saw the lack of both a large-scale convention centre and a modern rectangular stadium as gaps in Perth's range of facilities (Jones 2002). At the time of writing (July 2006) the convention centre has become a reality, situated along the city foreshore close to both public transport and freeway interchanges. However, debate is still ongoing regarding the function and siting of a state-of-the-art stadium, capable of accommodating 60,000 people at a high level of spectator comfort. *The West Australian* editorial (16 February 2006, 18) postulated that:

> The city has fallen well behind other State capitals which have built modern facilities designed to suit the major sports of football, cricket, rugby and soccer as well as other events. The WACA Ground and Subiaco Oval are a generation behind stadiums in other cities.

The editorial points to a generation gap, which has crept up on two of Perth's most iconic stadiums. Recent renovations have failed to dispel debates over the future of these two structures in an era that recognises the importance of the stadium as both a cultural and an urban/metropolitan icon. Rod Sheard, writing on the stadium as architecture for the new global culture, reflects upon the way in which each generation of stadia has 'raised the bar, adding a new level of sophistication and improved facilities (2005, 116). Sheard now points towards a fifth generation of stadia with the ability to play a role in the marketing and positioning of a global city. Ideally such a stadium would possess some elements of self-reflection grounded in the patina of heritage. Can it be that the city and state have lost the opportunity to transform Perth's third iconic stadium at Perry Lakes into a fifth generation structure capable of generating both the material and the mystical?

References

Australian Bureau of Statistics (ABS) *2001 Census of Population and Housing CDATA* (Canberra) official website accessed February 2006, available at; http://

www.abs.gov.au/Ausstats/abs@.nsf/d36c95a5d2ce6cedca257098008362c8/b7ad
1db8cb27192eca2570d90018bfb2!OpenDocumentwebsite.

Australian Commonwealth Games Association 'The Story of the British Empire & Commonwealth Games' official website accessed January 2006, available at: http://www.commonwealthgames.org.au/GamesInfo/General/General.htm

Bale, J. (1993) *Sport, Space and the City* (Routledge: London).

Crawford, G. (2004) *Consuming Sport: Fans, Sport and Culture* (London: Routledge).

Edmonds, J. (1962) *A Pictorial Record of the VIIth British Empire & Commonwealth Games* (Perth: Press, Publicity and Public Relations Committee for the Organising Council).

Edmonds, C. and Willmott N. (eds) (1962) *The Official History of the VIIth British Empire and Commonwealth Games, 1962, Perth-Australia* (Perth, Organising Council of the VIIth British Empire and Commonwealth Games).

Freestone, R. (1989) *Model Communities; The Garden City Movement in Australia* (Melbourne: Thomas Nelson).

Giulianotti, R. (1997) 'Supporters, Followers, Fans and Flaneurs: A Taxonomy of Spectator Identification in Football', *Journal of Sport and Social Issues* 26:1, 25–46.

Gregory, J. (2003) *City of Light: A History of Perth since the 1950s* (City of Perth).

Howard, P. (2003) *Heritage: Management, Interpretation, Identity* (London: Continuum).

Investigating Committee into the Development of Sport in Western Australia (1978), *The Development of Sport in Western Australia: Report of the Investigating Committee into the Development of Sport in Western Australia under the chairmanship of J. Bloomfield* (Perth: Community Recreation Council of Western Australia).

Jones, R. (1979) 'Local and Metropolitan Government', in Gentilli, J. (ed.) *Western Landscapes* (Nedlands: University of Western Australia Press).

Jones, R. (2002) 'Home and Away: the Grounding of New Football Teams in Perth, Western Australia', *The Australian Journal of Anthropology* 13:3, 270–282.

McDonald, C. (2004) *Perth Heritage and the 1962 Commonwealth Games Village* (unpublished Honours Thesis, School of Earth and Geographical Sciences, The University of Western Australia).

McKay, J. (1991) *No Pain, No Gain? Sport and Australian Culture* (New Jersey: Prentice Hall).

Perth Jaycee Publications of Organising Council (1962) *Official Guide to the VIIth British Empire and Commonwealth Games 1962* (Perth WA).

Real, M. (1998) 'MediaSport: Technology and the Commodification of Postmodern Sport' in Wenner, L. (ed.) *MediaSport* (London: Routledge).

Richards, D. (2003) 'The Acceptance of Modern Architecture in WA. The Architectural Competition for the Empire Games Village, City Beach 1961', in Lewi, H. and Neille, S. (eds) *Fading Events and Places: The Architecture of the VIIth British Empire & Commonwealth Games Village and Perry Lakes Stadium* (Perth: Department of Architecture, Curtin University of Technology).

Shaw, J. (ed.) (1984) *The Concise Encyclopedia of Australia* (Queensland: David Bateman).

Sheard, R. (2005) *The Stadium: Architecture for the New Global Culture* (Sydney: Pesaro).

Stannage, T. (1979) *The People of Perth* (Perth: Perth City Council).

Stickells, L. (2002) 'Modernism and the Games Village: Suburban Experimentation at the VIIth British Empire & Commonwealth Games, Perth 1962' in *Additions to Architectural History – XIXth Annual Conference of SAHANZ* (Brisbane: SAHANZ) 1, 16.

Syme, G.; Shaw, B.; Fenton, M. and Mueller, W. (eds.) (1989) *The Planning and Evaluation of Hallmark Events* (Aldershot: Avebury).

Town of Cambridge Municipal Heritage Inventory and Township Precinct Study (1997) prepared by Heritage and Conservation Professionals, Murray Street, Perth, June.

Town of Cambridge (2005) *Redevelopment of the Perry Lakes Stadium Site* (Business Plan Major Land Transaction, March.

Chapter 11

Port, Sport and Heritage: Fremantle's Unholy Trinity?

Roy Jones

Introduction

The City of Fremantle, with a population of c.25,000 is now but one municipality in the large and administratively fragmented Perth metropolitan area. For much of its history, however, it was separate from and, indeed, a rival settlement to the Western Australian capital. During the convict era of the mid nineteenth century, it was comparable in size to Perth and a local lobby group even attempted to have it designated as the colonial capital. While Perth-Fremantle rivalries remain (not least in sport) both their roles and their relative importance have since diverged considerably. Even so, Fremantle is currently Western Australia's largest port and the state's most popular tourist destination. It is also the only urban area in the state to have been seriously considered for World Heritage Listing. Although the City has a defined central conservation area and longstanding policies for heritage protection are in place, this application was ultimately not proceeded with because there was insufficient community/stakeholder agreement as to the desirability of such a move. This was in spite of the fact that the urban fabric of central Fremantle, and particularly of its West End area, conforms, in many ways, to the 'historic gem' archetype described by Ashworth and Tunbridge (2000, 155) in their classic work *The Tourist-Historic City: Retrospect and Prospect of Managing the Heritage City*:

> We label as 'historic gems' those, usually small, cities in which the historic resource is both so dramatic, extensive and complete and also so valued as to dominate their urban morphology, their identity and their policy options.

In Fremantle's case, the drama, size, completeness and value of its built heritage resources are not in question. However, this built heritage alone is unable to dominate the city's identity, morphology and policy options. These are also affected by Fremantle's other roles, notably as a major port and as a significant service, entertainment and tourist centre. It is the story of how these roles have become increasingly intertwined and, on occasion, contested over recent decades that forms the basis of this chapter. However, it is necessary to commence with some historical and heritage background information to indicate how this balance of interests came into being before the contestations between them can be explored in greater detail.

Historical and heritage background

Fremantle approximates closely enough to historic gem status for its evolution to be documented by reference to the four stage process (resource creation, dormant resource, race for survival, resource maintenance) of historic gem creation as outlined by Ashworth and Tunbridge (2000, 157) to be used as the basis for this section.

**Figure 11.1   Fremantle Prison (Fremantle Oval and 'Heritage' trees to the
                right)**

Photograph by the author.

Resource creation 1830–1910

In Ashworth and Tunbridge's idealised historic sequence, the heritage resource is conceptualised as being created in a single phase of economic and townscape development. However, Western Australia was such a remote and peripheral colonial outpost that sustained economic and demographic development did not eventuate until the turn of the twentieth century. Fremantle's built heritage resource was created in three separate phases over an 80 year period. When the Swan River Colony was first proclaimed in 1829, Fremantle was designated, by the Lieutenant Governor, James Stirling, as one its three inaugural townships and as the colony's port. However, the initial assessments of the area's economic and agricultural potential proved hopelessly optimistic (Cameron 1981), and the 'Swan River Mania' which had led to the colony's foundation rapidly subsided. 'The last major influx of settlers arrived in August 1830 by which time many of the earlier arrivals had already departed.' (Cameron 1979, 204). By 1848 the (non-Indigenous) population of the entire colony was 4,622, of whom only 426 resided in Fremantle (Shaw 1979, 334). The heritage resource from this period is therefore tied closely to the historical significance of the colonists' landing site and Fremantle's role as one of the initial European settlements in Western Australia, though the Round House (1831) occupies a prominent position and is one of the oldest buildings in the state.

The fortunes of the Colony and, still more so, of Fremantle were transformed between 1850 and 1868. During this period, Western Australia took over the role of convict transportation when this practice was discontinued by the eastern Australian colonies. Fremantle possessed an abundance of the local building material, limestone, and it was therefore selected as the site where the newly-arrived convicts would construct their own prison. This building was completed in 1855 and continued to serve as the colony's/the state's major gaol until the early 1990s. This building complex now ranks with Port Arthur in Tasmania as one of Australia's outstanding convict heritage sites (Figure 11.1). During this period, convict labour was also used to construct a number of notable and still extant buildings in the town including a Lunatic Asylum and the Warders' Quarters and, by 1859, Fremantle's population had risen to 2,392 (Shaw 1979, 336).

With the final demise of convict transportation, however, the colony reverted to slow growth until the final phase of (heritage) resource creation commenced in the 1880s with the onset of the Western Australian gold rushes. Between 1881 and 1901 the total population of Western Australia rose from 29,708 to 184,124. Fremantle was the gateway for this influx and its population likewise rose sixfold from 3,641 to 20,444. The uniform nature of much of Fremantle's townscape results from the building frenzy that occurred at this time and many public, commercial and residential buildings remain from this period. It was also the time when the port assumed its current nature and location. Initially, a rock bar blocked the mouth of the Swan River. A 'Long Jetty' was therefore constructed immediately south of the river mouth where goods could be offloaded from ocean going ships. Prior to the construction of the railway to Perth in the 1880s, the cargoes were then transported through the streets of the West End to river wharves for shipment upstream (see Figure 11.2). Under the direction of C.Y. O'Connor, the colony's Engineer-in-Chief, the rock bar was removed and Fremantle Inner Harbour was constructed in the lower reaches of the river. Victoria Quay and the Inner Harbour were completed in 1897 and, while it has considerable heritage significance, this is still very much a working and growing port.

Dormant resource 1910–1955

For much of the early and mid twentieth century Fremantle was essentially a dormant (heritage) resource. The gold rush subsided in the years prior to the First World War. The slow growth of the state in the early twentieth century was based on agricultural expansion, with export shipments distributed between Fremantle and a range of smaller regional ports. Although Fremantle became a major American and British submarine base in World War Two (Jones 1997), this did little to change the

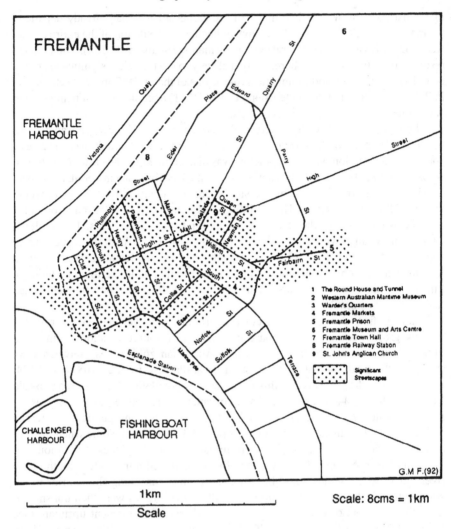

Figure 11.2   Map of Inner Fremantle

Map supplied by the author.

town's heritage base. Indeed, the convict-built lunatic asylum was used to provide accommodation for the American servicemen.

In the immediate post war period, port and port-related development tended to bypass the town. A new heavy industrial and port area was developed at Kwinana some 20 kilometres to the south and this has taken over much of the handling of bulk cargoes, such as wheat, oil and alumina. Local employment numbers declined as the wharves became more mechanised and, eventually, containerised and the town's commercial and retailing sectors faced increasing competition from newer, suburban shopping centres that were better adapted to Perth's increasingly car-dependant population (Marsh 1979).

Race for survival 1955–1970

In the 1960s and early 1970s Western Australia experienced its second mineral boom. Although this was mainly based on the immense reserves of iron ore in the Pilbara in the state's North West, as was the case with the gold rush that preceded it, it was Perth that benefited most in terms of population and jobs. Indeed, the metropolitan region was growing so fast that its population was (optimistically) predicted to double between 1970 and 1989 (MRPA 1970).

In these demographic circumstances, and in a state which had barely passed beyond the pioneering stage, economic development was prioritised over heritage preservation. Locally, however, Fremantle was favoured in built heritage terms by its subsidiary position in relation to Perth as a commercial centre and by the concentration of new port and heavy industrial development at Kwinana. A number of heritage battles were fought – and generally lost – in central Perth over this period, for example over reclamation of parts of the Swan River and over the loss of virtually all of the original Barracks to freeway development (Gregory 2003). In the mid-twentieth century, therefore the 'gold rush architecture' townscape of central Perth was largely replaced by one characterised by modern, high rise commercial buildings. Fremantle was seen by many, and not least by state planners, as rundown and obsolete by comparison and both the first metropolitan plan (Stephenson and Hepburn 1955) and an early City Council plan envisaged radical demolition and reconstruction. By the early 1970s, however: lessons from the destruction of built heritage in Perth had been learned and Fremantle's historic townscape had gained in local scarcity value; the first gentrifiers had begun to move into the town; and, particularly as Perth Airport began to replace Victoria Quay as Australia's 'Western Gateway' (Ewers 1971), the economic pressures for (re)development in Fremantle were less strong than was the case elsewhere in the metropolitan area.

Restoration of the former lunatic asylum from 1970 and its reopening as a museum and arts centre in 1972, the establishment of the heritage lobby group, the Fremantle Society in 1972 and the endorsement of heritage protection by the City Council in its landmark report 'Fremantle – Preservation and Change' (Fremantle City Council 1971) were all markers of this change in attitude. This shift was further entrenched when the Fremantle Society gained a majority on the City Council in 1977. The shift in the value set proved to be timely and became all the more important in 1983 when Fremantle became the venue for the 1986–7 Americas' Cup defence. This event massively increased redevelopment pressures on the town and brought the added variable of sport into the port and heritage equation for the first time.

Resource maintenance 1970 onwards

In recent decades the title of the City Council's 1971 report has proved particularly apt. The city has 'changed', both demographically, as gentrification has become increasingly entrenched, and in terms of employment, with productivist, and particularly port-related jobs decreasing while those in post-productivist, and particularly tourism-related, services have increased. But these changes have been largely dependant on the 'preservation' of Fremantle's pre-existing assets, notably the historic townscape, the port and the fishing boat harbour. These are the magnets which have attracted the newer elements of Fremantle's demographic and employment structure to the town.

In terms of resource maintenance, therefore, the government, the planners and the community of Fremantle have had to struggle to maintain a delicate balance between preservation and change. More than a decade ago the editors of this volume summed up this dilemma as follows:

> These recent changes to the nature of Fremantle have increased the potential for conflict in two ways. Firstly, Fremantle's new economic role as a centre for tourism in general and cultural tourism in particular, is not directly compatible with the city's original port and commercial functions. Secondly, many of the more recently arrived residents not only have a vision of Fremantle which differs markedly from that of the longer term inhabitants, they are also more likely to posses the economic, political and organisational abilities which will enable them to achieve their desired ends.' (Jones and Shaw 1992, 5)

On the first point, this chapter will document an ongoing series of heritage/port conflicts in this resource maintenance phase. On the second, however, and with specific reference to heritage/sport conflicts, the outcomes with respect to heritage resource maintenance have proved less predictable and have even led to novel questionings of the nature of Fremantle's heritage.

Heritage and the Port (Authority)

Fremantle and its port were essentially synonymous for a century and a half. A significant proportion of the local labour force was dependant upon the port, or on port-related activities, for their livelihoods. Over the resource maintenance phase of Fremantle's development however, both the spatial and the social divisions between the City and the port have come increasingly into focus.

In spatial terms, there is a jurisdictional split. The Fremantle Port Authority, a state government instrumentality, plans and manages its own development and its land is beyond Fremantle City Council's remit. In recent decades, most of the new port development in the Inner Harbour, notably the development of the container terminals, has taken place at North Quay, on the opposite bank of the river from Victoria Quay and the city centre. Nevertheless the efficient and evolving operations of a port which handles 90 per cent of the imports and 30 per cent of the exports of a rapidly growing state inevitably impact on the local community. The attitude of this community to the port over recent decades can best be described as ambivalent. Many in the local community would appear to possess an affection for the port and to exhibit in principle support for its retention, but numerous objections have been raised over the years to several of its externalities. Some of these relate to the nature of the shipping, such as vessels involved in the live sheep trade to the Middle East or American warships on goodwill or 'R and R' visits, some of which may be nuclear armed and/or powered (Fremantle City declares itself to be a nuclear free zone). Other conflicts arise over the needs of the port for goods transport access. A dispute over the Fremantle Eastern Bypass has continued for more than three decades. This was planned as an important link between the port and the metropolitan area's major road network. The first stage, including a new bridge over the Swan River was completed in the 1970s. Local objections to the second stage increased as the Fremantle suburbs through which it was planned to pass became increasingly gentrified. It is now likely that land purchased decades ago by the government along this route will be sold for residential development. Opposition, in the 1990s, to the expansion of rail access to the port was more directly heritage-related. The federal government planned to move freight interstate from Fremantle on kilometre long trains stacked two containers high. The rail route required for this encircles the historic West End (Figure 11.2) where such movements would have caused considerable disruption and noise pollution.

Similar conflicts also raged over (re)development within the Port Authority's boundaries. The objections to the expansion and relocation of fuel storage tanks by North Fremantle residents were, as with those over the Eastern Bypass, frequently based on amenity and property value concerns. However, when the Port Authority sought to demolish some early twentieth century wheat silos rendered obsolete by the movement of all wheat exports to a new terminal at Kwinana, they faced opposition from the Fremantle Society on the basis of their landmark significance and their historical role as a World War Two observation post.

To date, the Port Authority has generally succeeded in achieving its developmental ends within its boundaries, as with the demolition of the wheat silos on North Quay, but it has often been unable to influence planning decisions beyond them, as with the deletion of the second stage of the Fremantle Eastern Bypass from the Metropolitan Regional Scheme. Currently, however, port-heritage conflicts have focussed much more specifically on the Port Authority-city interface at Victoria Quay. In relative terms Victoria Quay has become less significant to the Port Authority over recent decades. More activity has shifted to North Quay with the growth of containerisation, and use of the Passenger Terminal declined significantly with the coming of wide bodied jet travel in the 1970s. However, the Port Authority (FPA 2000) still plans to use most of the Quay for shipping purposes for the foreseeable future. This includes some of the Authority's growth markets, such as cars and cruise ship passengers, and some of its more controversial shipping movements such as livestock and visiting naval vessels.

Nevertheless, the Authority sees Victoria Quay as becoming an increasingly mixed land use zone. The first stage of this development was extremely heritage-friendly. The spectacular Western Australian Maritime Museum (Figure 11.3) was opened at the west end of the Quay in 2002. This tall, sail-like building was designed to accommodate the Americas Cup-winning yacht, *Australia II*. Thus it links sport, port and heritage in a single structure. A strong personal link between people and the port is provided by its Welcome Walls, on which the families of migrants are invited to record the names of their relatives who arrived by ship in Fremantle. This port-heritage project has been so successful that it is now entering its third stage.

The contrasts with the Port Authority's current proposals for Victoria Quay could not be greater. In conjunction with finance company ING, the Port Authority now proposes to develop 25,000 square metres. of commercial space adjacent to Fremantle railway station (*The Sunday Times*, 16 April 16 2006). This proposal was immediately criticised by the Fremantle Society for a lack of public consultation and for its inappropriate scale and design, which included buildings up to eight storeys in height. Such a development would certainly not match the building character of the predominantly lower rise West End. It would also place a visual barrier between the town and the port. In particular it has the potential to block the iconic vistas whereby ships can currently be watched from the city centre as they pass by the end of several streets and dwarf the buildings as they do so.

Figure 11.3   Fremantle Maritime Museum and Victoria Quay

Photograph by the author.

Not surprisingly, therefore, a storm of protest greeted this proposal from the start. In early June an alternative proposal, with a height limit of four storeys, was proposed by the Fremantle mayor and the town's state and federal members of parliament. A week later, at a packed public meeting in Fremantle Town Hall, a vote: to reject the current development plan; to transfer control over the Port Authority land to the City Council; and to ensure early and frequent local consultation was passed with no dissent being recorded (*Fremantle Herald* 10 June 2006). ING is on record as saying that 'the final shape of the project would not be decided until public response to (the) design options was considered.' (Kelly 2006, 15). It is a further reflection of the strength of local feeling over this issue that the Fremantle Chamber of Commerce has been moved to place a statement on its website (http://www.fremantlechamber.com.au Accessed 24 October 2006) which supports the compromise suggestion of the mayor and the parliamentarians and underlines the importance of maintaining the 'view corridors' of the harbour along Market and Pakenham Streets but nevertheless re-emphasises the need for more modern, large scale commercial spaces in Fremantle if the number of local jobs and the diversity of the local economy is to be maintained.

Although negotiations are ongoing, this latest controversy would seem to presage a new phase in local port/heritage conflicts. Until recently, local community members and heritage organisations have sought to make their views known to the Port Authority and to lobby state and local council representatives and officials on various issues, such as the demolition of the wheat silos. The latest protest meeting demanded that the City, rather than a state instrumentality, be given planning control over the port. This follows two other recent requests that control over the port be moved to a very different level. In August 2005 the Fremantle Society lodged a request with the federal Department of Environment and Heritage for the placing of Fremantle Inner Harbour on the National Heritage List. This application was lodged in August 2005 and is still under consideration. Were the application to be successful the federal government would be able to provide what the Fremantle Society term 'proper protection' to the port and even federal grant monies to support aspirations for conservation (Fremantle Society Press Release 27 October 2005). A very different kind of federal intervention is advocated by contractor Len Buckeridge ('Boss urges Costello to Seize Ports' *The Australian* 4 July 2006, 6). Buckeridge, who 'plans to build his own port south of Fremantle' claimed that the operations of Fremantle port are 'appalling' and he called for federal control of the ports to improve their productivity. It is unlikely that his vision of federal control of the ports involves the kind of community consultation called for at the town hall meeting or, still less, the protection of heritage that the Fremantle Society seeks through the national listing.

In these circumstances, both the state government and the Port Authority find themselves caught between local demands for community consultation and heritage preservation on the one hand and federal and business demands for increased efficiency and improved throughput on the other. To date, this balance has largely been maintained. The port has catered for the growing traffic levels generated by the state's third mineral boom. It anticipates that this growth will see the inner harbour reach capacity in little more than a decade. An Inner and Outer Harbour Community Liaison Group was established in 1998. Cultural and commercial uses were brought onto Victoria Quay in and around the new Maritime Museum in 2002. A Buffer Definition Study was released in 2005 which claimed that '(a) critical component in implementing the buffer is integration of the study into local government planning processes and steps are being taken in this regard' (FPA website, accessed 24 October 2006). But, if the response of the June 2006 Town Hall meeting to the Port Authority's plans is any guide, these 'steps' need to be taken urgently. And, although the current situation is fraught, so much of Fremantle's identity and heritage is tied up with the port that it is to be hoped that these steps will be made in the right direction.

Sport(s) and Fremantle's heritage

The place of the port in Fremantle's heritage is undeniable. However two very different sports-heritage controversies have also occurred during Fremantle's resource maintenance phase. The first concerned the hosting, largely in and around the town's Fishing Boat Harbour, of the facilities, the boats, the crews and the spectators for the defence of the Americas Cup yachting trophy in the middle 1980s. The second, in the 1990s, revolved around the attempts by the (then) new Australian Football League team, the Fremantle Dockers, to establish its administrative and training facilities in the historic and partially heritage listed Fremantle Oval. While the impacts of the Americas' Cup defence on Fremantle's townscape were far more fundamental, the controversies over the Dockers' headquarters clearly showed that, for many in the local community, 'heritage' had a meaning that extended well beyond the built environment.

The Americas Cup defence

An interesting concatenation of circumstances occurred in 1983. In Newport, Rhode Island, *Australia II* broke the New York Yacht Club's more than a century long stranglehold on the Americas Cup, thus ensuring that the 1986–87 Cup defence would take place off Fremantle. In Australia, not only did the Australian Labor Party triumph in both the federal and Western Australian elections, but the state and federal members for Fremantle became Treasurers in their respective governments. This sporting victory on the other side of the planet triggered an exceptional burst of investment in the redevelopment of Fremantle. The election victories meant that federal, state and local government would work together to ensure that this development boom would proceed with as little intergovernmental conflict and, indeed, as much intergovernmental support as possible.

There is no doubt that this sporting victory unleashed an investment surge in Fremantle on a scale that had not been seen since the gold rush a century earlier. An estimated $2.8 billion (*Western Mail Magazine* 27 September 1986, 23) was spent in preparations for the Cup defence, some on the immediate event requirements, but much invested in the hope of returns from longer term economic and tourist growth. Clearly the potential was there for this far-off sporting victory to have a massive impact on Fremantle's historic built fabric, particularly if those whom Shaw (1989, 39) termed the 'phallicists' gained permission to develop high-rise structures, such as Alan Bond's cup-related Observation City hotel, which now towers above the suburbia of Scarborough Beach a few kilometres to the north.

This did not happen. The new (*Challenger*) yacht harbour (Figure 11.2) , constructed adjacent to the existing Fishing Boat Harbour, was by far the largest local building project in physical terms. On land, the extension to the Esplanade Hotel was the only significant example of a cup-related building that was out of scale with the existing Fremantle streetscape. In terms of architectural manners, even this blends in far more with the town's historic buildings than do the modernist Princess (Diana) of Wales Wing of the Fremantle Hospital – adjacent to the historic prison – or some high rise 1960s flats in the centre of the town. The Americas Cup-fuelled redevelopment largely took the form of the repair and restoration, and often the conversion (e.g. of warehouses to luxury apartments), of existing structures. In the historic West End alone the number of planning applications for work of this type increased from 105 in 1983 to 251 in 1986 (Fremantle City Council Minutes,

16 March 1987). Indeed some of the most valuable local investment was publicly funded and, if not invisible, it was largely unobtrusive in the form of refurbished water pipes, sewers and cables, improved roads, cycle ways and lighting, restored dunes and expanded car parking.

That such a comprehensive and integrated programme of improvements could be implemented so rapidly resulted from two eventualities (Jones and Selwood 1991). Firstly, the political/electoral events of 1983 ensured that an intergovernmental committee directed by the Mayor of Fremantle and the federal and state Treasurers made funding for these projects readily available and that intergovernmental rivalries were minimised. Secondly, the value shifts associated with the 'race for survival' of Fremantle's built heritage in the 1970s had enabled both state and local governments to identify a number of strategies and projects to protect and enhance Fremantle's townscape. In 1979, the State's Metropolitan Regional Planning Authority had devised a list of eight proposed 'actions' to promote Fremantle's role as a recreational and cultural centre and the City Council's 'Fremantle in the Year 2000' (1980) also documented a number of (what were then seen as) longer term development projects. These included expansion of the marina/fishing boat harbour and foreshore and hotel developments. At the time that these proposals were put forward, the possibility that the 'Auld Mug' would be prised from the New York Yacht Club and that over $80 million of state and federal government grants would be available for the implementation of these projects was not even anticipated. Nevertheless the presence of these blueprints was vital to the rapid completion of many cup-related environmental improvements and tourist and heritage developments.

A final and significant sporting-heritage point is that Fremantle lost the Americas Cup in 1987. Hall and Selwood (1989) argue that the city may not have been able to withstand the development pressures of more than one defence and that this was a case of 'Cup Lost: Paradise Retained.' Certainly pro-development voices saw 'the initial experience (as) a good guide to big and small entrepreneurs in the future as to where to invest their risk capital for the next cup challenge without government interference' (Dale 1986, 185). Fortunately for (heritage) resource maintenance, the cup was lost, but government interference in the form of conservation controls over Fremantle's historic townscape remains.

Nevertheless, while the single Americas Cup defence was crucial in terms of maintaining – and, indeed, repairing, restoring and refurbishing – much of Fremantle's built fabric, its role in changing the town's social fabric was immense. For more than three years, Fremantle was associated with a millionaire's sport and with the international 'jet set'. Such a juxtaposition inevitably speeded up a gentrification process that was already proceeding apace. While this undoubtedly brought into the town a population that was increasingly willing and able to defend and preserve the town's built heritage – in which many of them had invested as homebuyers – they did not necessarily conceive of local heritage(s) in wider terms. Their ambivalence over Fremantle's role as a port has already been discussed. Their attitudes to Fremantle's traditional character were to be demonstrated again when, after losing a yachting trophy, Fremantle gained a major Australian Rules football club.

Docking the Dockers

As a side effect of the explosion of the Western Australian population in the gold rush of the 1880s and 1890s, interstate migrants from Victoria popularised the playing of Australian Rules football in Western Australia. Fremantle clubs dominated the first local competitions in the late nineteenth and early twentieth centuries and both the South and East Fremantle Football Clubs played Western Australian Football League (WAFL) matches at Fremantle Oval until East Fremantle built a new ground in that suburb following World War Two. Fremantle Oval is in the centre of the town, between the markets and the prison and its imposing Victoria Grandstand (Figure 11.4), opened in 1897 is, like these neighbouring buildings, heritage listed.

Until the 1980s, the Western Australian Football League was the state's premier sporting competition. It inspired strong (sub)urban identities and rivalries. As one of the most successful teams of the post World War Two era, South Fremantle had a high profile in the town and triumphal parades were held through the city streets whenever South Fremantle won 'the flag'. In 1987, however, not only did Fremantle lose the Americas Cup, but a state team, the West Coast Eagles, entered the Victorian Football League competition and, almost immediately, the WAFL came to be seen as merely a second tier, 'feeder' competition. This was part of a more general late twentieth century move by the football codes and other team sports to establish national leagues in Australia (Jones 2002). Although the traditional Foundation Day 'derby' match between East and South at Fremantle Oval is still important locally, crowds at WAFL matches have declined significantly over the last two decades and the Victorian Football League (renamed the Australian Football League in 1990) has become the nation's major sporting competition. Just as a proportion of Fremantle's shipping traffic has now moved out of the Inner Harbour, so too have many of the town's football fans deserted Fremantle Oval to watch Australian Football League matches 15 kilometres away at the state's main ground, Subiaco Oval.

In 1993, the AFL carried its strategy of 'nationalising' the league to another level by creating a second Western Australian team. Although the West Coast Eagles had tried to position themselves as essentially 'placeless' in order to receive state wide support, Subiaco Oval is very close to the Perth CBD and the Eagles were inevitably seen as a (and indeed, up to that time, the) Perth team. For marketing reasons it was therefore decided that, given Fremantle's strong football heritage and the traditional Perth/Fremantle and North of the (Swan) River/South of the River rivalries in the metropolitan area, the second team should have a Fremantle identity. In 1994, the Fremantle Dockers name and club colours were unveiled. The club had its first training session (at Fremantle Oval) in October of that year and, in 1995, it entered the national competition.

In symbolic terms, the Dockers were closely linked to Fremantle from their inception. Even though 'dockers' is a non-Australian term (Australian ports employ waterside workers or 'wharfies'), the club's maritime link is clear and is reinforced by the anchor logo on their shirts. Furthermore, their colours include red, white and green acknowledging the large local Italian population which is strongly represented in the Fremantle fishing industry, further extending the sport, port and heritage links. In terms of time as well as place, the club has also sought to invent a tradition (Hobsbawm and Ranger 1980) that they do not, strictly, possess and to forge symbolic links with the famous Fremantle teams of the early twentieth century. One aspect of this is to highlight the importance of their Anzac Day match and a local Gallipoli veteran, wearing a Dockers scarf, was photographed with the club captain on Anzac Day 1998 (*The West Australian*, 25 April, 1998). The team also wears a red and white strip based on the original Fremantle club's design in the AFL's 'Heritage Round' of matches, colours which were also adopted/inherited by the WAFL's South Fremantle, who have had their home at Fremantle Oval for more than a century.

However, it has proved more difficult for the Dockers to progress from a symbolic to a physical presence in Fremantle. Fremantle Oval is too small and lacks the seating, floodlighting and parking capacity to stage AFL games. All AFL games in Western Australia are therefore played at the Western Australian Football Commission managed Subiaco Oval. From the start, however, the Dockers sought to strengthen its local links by establishing its administrative and training buildings at Fremantle Oval. This proposal was strongly supported by the AFL, the Mayor of Fremantle and, if several opinion polls were any indication, the majority of the local community. Nevertheless, this project experienced a number of delays and setbacks, with opposition to the proposal coming from at least three quarters.

The first two sets of objections, from (very) local residents concerned abut noise and disruption, and from South Fremantle Football Club, which would have to share the Oval with an upstart, but now more powerful and wealthy, organisation, were both predictable and were rapidly and successfully resolved. The third set of objectors were not only the strongest and most determined, they were also the most heritage-related. A group of objectors, largely associated with the Fremantle Society, attacked the building plans in detail over their visual impact on the vista from the town to the gate house of the prison and over the planned removal of a heritage listed Moreton Bay fig tree (*Fremantle Gazette*, 9 August 1994) (Figure 11.1). These objections were raised through Fremantle City Council's Planning Committee. At first, they received relatively little attention. But, after the Dockers had modified their building plans several times to meet the Society's and other councillors' concerns and still failed to receive Council planning approval, the club threatened to move its headquarters elsewhere in the metropolitan region and to retain only a nominal link with Fremantle.

This threat had an immediate effect. It received headline treatment in the local press. Irate letter writers contended that football was more important to Fremantle's heritage than were trees or vistas. A large protest march was organised (*Fremantle Gazette*, 8–9 August 1997) at which many banners referred to 'keeping' the Dockers in Fremantle, a powerful reflection of the success of the symbolic campaign to identify the club with the town over a mere two year period. Very shortly after these protests, an architectural/planning compromise was reached, building commenced and the Dockers' clubhouse, training and administrative headquarters were opened at Fremantle Oval in October 2000 with some fanfare, but remarkably few contemporary or subsequent protests.

Since then the links between club and town have grown. Fans are invited to watch training sessions at the Oval on Monday evenings, when patronage at the many local cafes, restaurants and bars is relatively low and city centre parking is readily available. In a town well supplied with pubs as part of its port heritage, there are many Fremantle venues where Dockers matches can be watched on big television screens.

Although the team's on field record has generally been indifferent, the 2006 season brought its first finals win and the many of the cafes, restaurants and pubs which have characterised the town's 'cappuccino strip' and its environs since the Americas Cup defence were the venue for fans' celebrations. Indeed, this close identification between the team and the town has invited atypically favourable comparisons with the (far more successful) West Coast Eagles who lack a geographical focus for the celebration of their more numerous sporting successes.

Finally, some material and townscape recognition has been given to the place of the game (rather than the buildings) of football in Fremantle's heritage. Prior to the East Fremantle/South Fremantle derby match on Foundation Day 2006, the Premier of the State and the Mayor of Fremantle unveiled a statue on a roundabout at the gates of Fremantle Oval (Figure 11.4). The statue was based on a classic photograph depicting a spectacular mark taken by a South Fremantle player over his East Fremantle opponent in the 1956 preliminary final. To date, at least, the town appears to have achieved predominantly successful resolutions of its sport-heritage conflicts. At a time when controversy over the proposed expansion of Subiaco Oval is growing, Fremantle may well be relieved that the state's major stadium, unlike its major port, is some distance from its own heritage precincts.

Figure 11.4   'Mark' statue, between the Victoria Grandstand and Fremantle markets

Photograph by the author.

Conclusion

Australia is a settler society and it was largely populated through its gateway primate capital cities. In these circumstances, it is hardly surprising that these points of entry are, simultaneously, heritage epicentres and physically dynamic localities within their respective metropolitan regions. In recent decades, the potent combination of central locations, historic buildings and waterfront proximity have attracted gentrifiers to such sites, adding social dynamism and diversity into their already complex land use mixes. Various elements of the heritage and planning issues facing Fremantle can be discerned in the port areas of other state capitals. Port Adelaide is currently deindustrialising, gentrifying and losing its downmarket 'Port Misery' identity (Oakley 2005: Rofe and Oakley 2006). Melbourne's Docklands redevelopment incorporates a major sports stadium and former port areas in Darling Harbour in Sydney have been redeveloped as a recreational and, to a lesser extent, a heritage precinct (Waitt and McGuirk 1996; Shaw this volume).

In all these cases there is 'a tendency... to design them as places which filter ... experiences rather than integrate them with the quotidian' (Clelland 2000, 105). Residents, tourists and investors in these historic port precincts are seeking a lifestyle product which provides them with elements of the economic reality of the operation of a port, or of the social reality of an established, and usually working class, community involved with a local sporting team. But many wish their experiences of these places to be 'filtered', as Clelland puts it, or to be a 'hyperreality' (Eco, 1986) in which they are protected from the smell of sheep ships, the noise, bustle and disruption of sports fans or the obtrusiveness of heavy transport systems serving the docks. There is no doubt that these new residents, visitors and investors, are evidence that 'heritage is an economic instrument in policies of regional and urban development and regeneration' (Graham et al. 2000, 256) Certainly this is true in Fremantle's case where, in the face of significant employment decline on and related to the docks, heritage (aided, somewhat fortuitously, by an Americas Cup defence) was instrumental in providing the town with an alternative economic and demographic base.

Where heritage-related uncertainties and disagreements have arisen, these frequently revolved around conflicts over the extent to which Fremantle's heritage was perceived as being a preserved visual or surface feature or as an ongoing and, therefore, both a changing and an unfiltered quotidian experience. The former might be represented by an abandoned warehouse which has been gutted behind its historic facade and turned into luxury apartments and the latter by the trucks passing through the town carrying live sheep en route to the Middle East. Over more than thirty years of 'resource maintenance', Fremantle has been relatively successful in maintaining a balance between these extremes. Both the 'heritage lobby' (and, notably, the Fremantle Society) and the developers (and, notably, the Port Authority) have had their wins and their losses in various disputes. But, over this period the amount of cargo handled by the port, the number of tourists visiting the town, the prices of Fremantle real estate in comparison with those for the rest of the Perth metropolitan area and even the level of Dockers memberships in comparison with those of the other AFL clubs have all shown upward trends. In November 2006, the city hosted, and presented Fremantle as an exemplar at, an Australian ICOMOS conference on 'Challenge and Change in Ports, their Towns and Cities.' One speaker at the conference controversially raised the prospect of removing the port from the Inner harbour and covering North Quay with luxury housing but, while this proposal 'sent shock waves through his audience' ('Kill Port for Homes: Prof' *Fremantle Herald* 18 November 2006) it is at odds with the balance that has been maintained in Fremantle of late. Over three decades of heritage resource maintenance, the evidence of the four growth trends mentioned above would indicate that Fremantle has indeed been rising to the challenge of its changes.

References

Ashworth, G. and Tunbridge, J. (2000) *The Tourist-Historic City: Retrospect and Prospect of Managing the Heritage City* (Oxford: Elsevier Science).

Cameron, J. (1979) 'Patterns on the Land, 1829–1850', in Gentilli (ed.).

Cameron, J., (1981) *Ambition's Fire; the Agricultural Colonization of Pre-Convict Western Australia* (Nedlands: University of Western Australia Press).

Clelland, D. (2000) 'Tourism and Future Heritage: Genes – Dreams – Designs, or the Engaging of the Syndrome of S, S and S', *Built Environment* 28:2, 99–116.

Dale, A. (1986) 'In Quest of the Holy Grail: W. A. and the America's Cup', in O'Brien (ed.).

Eco, U. (1986) *Travels in Hyperreality: Essays* (San Diego: Harcourt Brace Jovanovich).

Ewers, J.K. (1971) *The Western Gateway: A History of Fremantle* (Nedlands: University of Western Australia Press).

Fremantle City Council (1971) *Fremantle: Preservation and Change.*

Fremantle City Council (1980) *Fremantle in the Year 2000.*

Fremantle Port Authority (2000) *Port Development Plan.*

Gentilli, J. (ed.) (1979) *Western Landscapes* (Nedlands: University of Western Australia Press).

Graham, B., Ashworth, G. and Tunbridge, J. (2000) *A Geography of Heritage: Power, Culture and Economy.* (London: Arnold).

Gregory, J. (2003) *City of Light: A History of Perth Since the 1950s* (Perth: City of Perth).

Hall, C. and Selwood, J. (1989) 'Americas Cup Lost: Paradise Retained? The Dynamics of a Hallmark Event', in Syme et al (eds).

Hobsbawm, E. and Ranger, T. (1983) *The Invention of Tradition* (Cambridge: Cambridge University Press).

Jones, R. (1997) 'Geostrategy and Geotactics: Recent Naval Port Developments at Garden Island, Western Australia', *Marine Policy* 21:4, 363–376.

Jones, R. (2002) 'Home and Away: the Grounding of New Football Teams in Perth, Western Australia', *The Australian Journal of Anthropology* 13:3, 270–282.

Jones, R. and Selwood, J. (1991) 'Fallout from a Hallmark Event: Fremantle after the Departure of the Americas Cup', in Royal Australian Institute of Parks and Recreation (ed).

Kelly, J. (2006) 'Waterfront Dispute' *The Sunday Times,* 16 April, Prestige Property 5.

Marsh, C. (1979) 'Retailing', in Gentilli (ed.).

Metropolitan Regional Planning Authority (1970) *The Corridor Plan for Perth.* (Perth: Metropolitan Regional Planning Authority).

Oakley, S. (2005) 'Working Port or Lifestyle Port? A Preliminary Analysis of the Port Adelaide Waterfront Development', *Geographical Research* 43:3, 319–326.

O'Brien, P. (ed.) (1986) *The Burke Ambush: Corporatism and Society in Western Australia* (Nedlands: Apollo Press).

Rofe, M. and Oakley, S. (2006) 'Constructing the Port: External Perceptions and Interventions in the Making of Place in Port Adelaide, South Australia', *Geographical Research* 44:3, 272–284.

Royal Australian Institute of Parks and Recreation (ed.) (1991) W*ho Dares Wins – Parks Recreation and Tourism: Conference Proceedings* (Melbourne: Royal Australian Institute of Parks and Recreation).

Shaw, B. (1979) 'The Evolution of Fremantle', in Gentilli (ed.).

Shaw, B. (1989) 'Fremantle and the America's Cup ... The Spectre of Development', *Urban Policy and Research* 3:2, 38–40.

Shaw, B. and Jones, R. (1992) *Historic Port Cities of the Indian Ocean Littoral: the Resolution of Planning Conflicts and the Development of a Tourism Resource.* Indian Ocean Centre for Peace Studies, Occasional Paper 22.

Stephenson, G. and Hepburn, J. (1955) *Plan for the Metropolitan Region, Perth and Fremantle, WA.* (Perth: Government Printer).

Syme, G., Shaw, B., Fenton, M. and Mueller, W. (eds) (1989) T*he Planning and Evaluation of Hallmark Events* (Aldershot: Avebury).

Waitt, G. and McGuirk, P. (1996) 'Marking Time: Tourism and Heritage Representation at Millers Point, Sydney', *Australian Geographer* 27:1, 11–29.

Chapter 12

Places Worth Keeping

Rosemary Rosario

Places of cultural significance enrich people's lives, often providing a deep and inspirational sense of connection to community and landscape, to past lived experiences. They are historical records, that are important as tangible expressions of Australian identity and experiences. (Burra Charter).

Whose heritage? Subiaco 2002

On 24 July 2002, Winthrop Hall, the main auditorium at The University of Western Australia, was packed with over one thousand people for a Special Electors' Meeting called by the City of Subiaco to address the overwhelming negative public response to the release of the review of its municipal inventory. The draft inventory had been released in March for the purpose of public consultation. However, assisted by the emergence of a community group calling itself 'Heritage Gone Mad', the consultation that ensued had gone seriously wrong. As the Subiaco council's heritage consultant for the project, one could only feel dismay as one after another angry, often ill-informed, speakers denounced the idea of protecting the 'so-called heritage' of this early-twentieth-century inner suburb of Perth. How had things gone so horribly wrong? How could we salvage something and move on? And, for the longer term, the question of how this would impact on the wider context of heritage in Western Australia loomed large. For that night, however, the questions were far more immediate.

This chapter considers the case for the protection of heritage at the local level in the older suburbs of Perth, Western Australia, as seen from the perspective of a heritage professional with over fifteen years of practical experience. It also reviews some of the issues that predated the Subiaco events, looks at their immediate aftermath, and then considers some of their longer-term consequences. The discussion investigates some of the reasons why these protests occurred in Subiaco, but not in other areas, and why they occurred at this particular time. In conclusion, ways in which local heritage can be managed successfully, and factors that can contribute to that success, are given due consideration.[1]

---

1   The author gratefully acknowledges the contribution of her colleagues; town planners Chris Antill, Murray Castleton and Martin Richardson; Michael Betham and Stephen Carrick of the Heritage Council of Western Australia (HCWA), and Subiaco real estate agent Michael Hoad, all of whom generously contributed time and ideas to give broader perspectives to the issues discussed in this chapter.

The events in Subiaco did not happen in isolation. Their wider context included not only the attitudes of the immediate community affected by the release of the Subiaco municipal inventory review, but also a much broader spectrum of issues. These included state and local government roles in the protection of heritage, the impact of heritage on a volatile property market, and the perceived impact of heritage on individual property owners' rights. The events were also a result of the unique history of development of Perth, and of the ways in which the people of Perth, its suburbs and surrounding areas, have perceived themselves, their history and their environment over that time.

Residential development in metropolitan Perth

Settlement of the Swan River Colony began in 1829 with initial centres being established at the port of Fremantle at the mouth of the Swan River, the capital Perth 15 kilometres up river at the base of Mount Eliza, and the proposed agricultural centre at Guildford approximately 12 kilometres further east at the junction of the Swan and Helena Rivers. In 1830, establishment of a centre at Kelmscott, roughly 25 kilometres southeast of Perth on the Canning River, completed the picture. During the nineteenth century, trade clustered around these nodes, with the river as the early transport link (Burke, M. 1987; Pitt Morison and White 1979).

The emergence of inner suburbs

Discoveries of gold in Western Australia, from 1883, resulted in a rapid increase in population growth and urban expansion. The 1880 recorded population of 29,561, rose to 48,502 in 1890, and to 184,124 by the 1901 census (Sondalini no date). By the late-nineteenth century land close to Perth was subdivided for residential development and a ring of suburbs had developed around the city. Areas including North Perth, Leederville and Subiaco were part of this expansion that provided mainly working class accommodation in small detached houses in the new residential subdivisions. Roads in the new subdivisions followed a grid pattern with lots of varying sizes, often approximately 400 to 600 square metres (Stannage 1979).

Further subdivision of areas such as Nedlands, Dalkeith and Claremont, on the northern banks of the Swan River, and in the beachside suburb of Cottesloe, saw the suburban development of the western suburbs between Perth and Fremantle, with larger homes on larger lots. Lots were often around 1,000 square metres, or the traditional quarter acre, many with ocean or river views. Together with larger homes, areas of workers' cottages on smaller lots had developed around the railway line that opened between Fremantle and Guildford in 1881. This development extended from Cottesloe and Claremont, through Subiaco and Leederville to the city of Perth and beyond (Selwood 1979).

The urban geography of Perth over the twentieth century responded to changes in the state's economy that were linked to both local and world affairs. Periods of suburban expansion followed World War I as service personnel returned home. At the same time population increases resulted from increased migration under various government assisted settlement programs. In the inter-war period suburban development extended out from Perth's core of inner suburbs to areas such as Wembley and Floreat Park (Rosario 1993), and by the 1950s suburban settlement had extended to the coast at City Beach (see Figure 12.1).

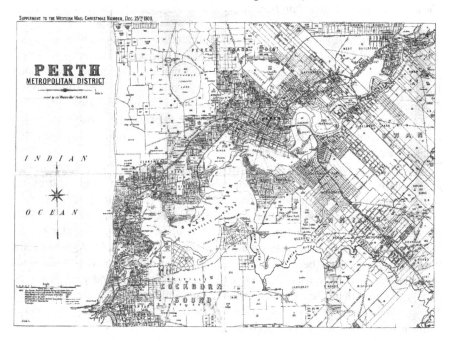

Figure 12.1   Map of metropolitan Perth showing extent of development by 1909

Source: Battye Library, Perth, Western Australia: items approved 466C (37/13).

The date of development of the various suburban areas is clearly reflected in the way these were both planned and developed. The road layout, subdivision pattern, house styles and landscaping of public areas and private gardens all contribute to a picture of development that physically reflects both a period of history and the way of life of the people who lived there. The areas closest to the city of Perth were developed on a grid pattern of street layout and houses were developed mainly in the Federation to early inter-war periods. As development expanded to areas further from the city centre in the 1920s and 1930s, the principles of garden suburb development resulted in curvilinear street patterns around areas of public open space or amenities. House styles changed, reflecting changing lifestyles and the move to private car ownership requiring garages (Freestone 1989).

While it is generally true that residential areas reflect their period of development, and that change is inevitable, it is also the case that some areas develop in response to specific significant events at a particular time that are historically important. In such areas the housing stock that remains has heritage value because it reflects both the specific time and the way of life of that era. The area becomes significant for those identified heritage values taken as a whole, rather than for the values of individual houses.

The changing face of suburban Perth

From the 1960s the character of the Perth metropolitan area began to change as a result of internal factors including the nickel and iron ore boom, and external factors such as changes to the architecture and design of the suburban home (see Chapter 10). During the 1960s and 1970s period homes in many of the older suburbs were demolished or modernised in a way that removed almost all their original features, such as decorative stucco work, iron lace or timber fretwork. This was a low period for those who loved the style and decadence of the older Federation era homes, built around the turn of the twentieth century.

It is often the legacy of works carried out in the 1960s and 1970s that results in arguments for demolition of an old home because it has lost its original detailing and is described as 'not intact'. However, if an area is significant as a whole it is important to encourage retention of those places that may not be individually significant, but that add to the significance of the area as a whole. This idea has not been well understood, as the inner suburbs of Perth, characterized by a concentration of heritage places, have been exposed to pressures for redevelopment. While some local planning schemes have endeavoured to protect places of individual significance, there has been little evidence of protection of the significance of an area as a whole. This applied during the period of redevelopment of the 1960s and 1970s, and in many areas has continued to the present day.

Heritage movements in Perth's older suburbs before 1990

The National Trust of Australia (WA), established by an Act of Parliament in 1959, was often the lone voice against urban redevelopment that characterised the 1960s, 1970s and 1980s. Moreover, even if the National Trust classified a building, this did not imply any statutory protection against demolition or insensitive re-development.

By the latter decades of the twentieth century, increased concern over the extent of demolition of homes representing the state's early history, led to the emergence in some communities of a movement to retain and conserve heritage buildings. Groups that emerged included the Fremantle Society, the Mount Lawley Society and the Guildford Association. It is no accident of history that in these three communities the record for successful heritage conservation has been considerably better than in some other areas.

In Subiaco the earliest attempt at heritage identification occurred in 1985 when the Council, under the leadership of pro-heritage mayor Richard Diggins, commissioned architect Ian Molyneux to carry out a heritage survey of the area. The ensuing report, *Survey of the National Estate in Subiaco* (Moyneux 1985), known colloquially as the 'Molyneux Report', recommended a number of buildings for notification to the National Estate and for nominating to the National Trust of Australia (WA) for potential classification. In addition over 1,000 buildings were recommended for inclusion in the City of Subiaco's Town Planning Scheme 3. Even more significantly the report identified the character of Subiaco as rare if not unique in the metropolitan area, with the exception of Fremantle and North Fremantle, in that it included the whole of the municipality (see Figures 12.2 to 12.5). Even though the report acknowledged that 'Change and renewal in a living community such as Subiaco are inevitable' (*ibid.* Vol.1 p. 3) it led to controversy across the community and the main recommendations were never adopted. Although the National Trust classified a few buildings and a precinct in the area known as 'the Triangle', no statutory heritage protection was implemented and no further action was taken for a number of years.

Figure 12.2   A worker's cottage in Subiaco

Photograph by the author.

The Heritage of Western Australia Act 1990

Heritage movements at grassroots level were by no means common and by 1990 when the *Heritage of Western Australia Act* finally gave some statutory protection to heritage, only a handful of local authorities had any formalised policies for heritage protection. Those that had addressed the issue included the City of Fremantle, which had protected heritage through its town planning scheme since the 1970s; the City of Swan that had scheme provisions for the protection of heritage precincts dating from 1989; and the Town of Claremont that had adopted a heritage list into its scheme in the 1980s. By

Figure 12.3   Bagot Road, Subiaco streetscape

Photograph by the author.

contrast, recommendations for inclusion of places in the City of Subiaco's town planning scheme through the 'Molyneux Report' in the 1980s had not been carried out.

In Western Australia there was no legislative protection for heritage before 1990. *The Heritage of Western Australia Act* (the Act) established the Heritage Council of Western Australia and made provision for a register of places of cultural heritage significance that would be protected by law. The Act included provisions for the entry of individual places or precincts into the Register of Heritage Places and for their protection. The Act was mainly concerned with individual places of state heritage significance, and not with issues of local heritage.

Municipal inventories

The Heritage Act did include one section that applied to local heritage matters. This was the requirement for all local authorities in WA to prepare inventories of buildings within their districts that were in their opinion of cultural heritage significance. (Heritage of Western Australia Act 1990, Section 45). The lists were required to be prepared within four years of the Act coming into being, and to be reviewed every four years thereafter. There were no legislative requirements for the protection or management of places on the lists either under the heritage legislation, or any other statutory instrument.

The provisions of the Act in respect to municipal lists, or municipal inventories as they came to be known, were in fact quite vague. There was no standard process to be followed for their compilation, and no recommendations as to what was to be done with municipal inventories when they were completed. The Heritage Council initiated the process in the early 1990s by the infusion of a small sum of money ($3,500) per local authority. The local authorities were then expected to add a comparable sum, and for this they were expected to commence the process. The Heritage Council provided guidelines for the work and this formed the basis of the brief to heritage consultants who were generally engaged to carry out the work. The Act also required that the inventories be 'compiled with proper public consultation' (*Heritage of Western Australia Act* 1990, Section 45), but the detail of this was up to the discretion of the individual local authority. There was no requirement to consult with owners and no requirement to protect in any way the places identified for inclusion on municipal inventories.

In the early 1990s the heritage practitioner's role often involved outlining this 'new' concept of heritage to local governments and their communities. The method employed often involved the appointment of a reference group comprising council staff, interested councillors and community members to assist with drawing up the list. There was considerable interest and support for the process, although this varied depending on the area.

During this period, one of the challenges of working on early municipal inventories and in the heritage area in general, involved exploring the meaning of

Figure 12.4   Federation Queen Anne style house, Subiaco

Photograph by the author.

Figure 12.5   Streetscape of workers' cottages, Shenton Park

Photograph by the author.

heritage. Before this, heritage had often been thought of in quite narrow terms. Old buildings had been considered as heritage places largely on account of their visual appearance. The heritage legislation however defined heritage more broadly in terms of aesthetic, historic, social and scientific values. This led to the consideration of a broader range of places as having heritage value. For communities, to appreciate that heritage places extended beyond the concept of the stately home, opened up a range of opportunities. Some councils began to appreciate a much broader range of heritage places and included parks, bridges, monuments and the like, as well as quite modest buildings, on their municipal inventories. However a major limitation of the heritage legislation as it applies to municipal inventories, is the focus on buildings. Although many councils expanded their thinking to include a range of places other than buildings, there was no requirement for them to do so.

Despite the compilation of lists that varied in terms of quality and content, by the mid-1990s the overall outcome of the inventory process had resulted in an increase in awareness of heritage issues. A few local authorities went on to include some, or all, of the places identified in heritage schedules attached to their town planning schemes. In such cases the requirement to consult with owners became imperative and it often required considerable political will for councillors to commit to this level of heritage protection. Generally, protection through the town planning schemes has only occurred in areas where there has been considerable community support for heritage and where extensive consultation with owners has been carried out. The level of consultation for protection though planning schemes being far more extensive than the requirement for consultation intended under Section 45 of the heritage legislation.

While local authorities were charged with preparing municipal inventories, the process of protecting heritage areas or precincts was generally neglected. While provision for precincts to be entered into the Heritage Council's Register of Heritage Places is included in the Act, this has proved administratively complex. There are no requirements under the Act for local councils to protect precincts or heritage areas, which is unfortunate as their protection though a town planning scheme is much more appropriate than through the provisions of the Act. Furthermore, while the inventory process might have been appropriate for councils with a few iconic heritage places, the process was not well suited to places where whole areas or streetscapes had heritage value. The outcome for councils such as Subiaco, was that while some individual buildings were listed, numerous, possibly equally significant, places were omitted.

The lack of a clear direction for local authorities regarding the management of places on their municipal inventories led to uncertainty and confusion for both the councils themselves and the owners of listed places. While some councils were keen to protect the places they had identified through the inventory process, the majority were more inclined to consider their obligation fulfilled with the preparation of the list. A major reason for this was the requirement for a further process of consultation with owners of the listed places, but the overriding reason was lack of any statutory requirement for councils to do more than compile the list.

The Act required that a copy of each local authority's municipal inventory be lodged with the Heritage Council. This did not imply that the Heritage Council had adopted or endorsed the inventory in any way, or had any role in its management, or in any other local heritage issues. While the Heritage Council maintained that the lodgment of the list was their only involvement, some owners of properties listed on a municipal inventory were suspicious that this could lead to their properties appearing on the Register of Heritage Places without further consultation. While this belief was erroneous, the ambiguity of the situation regarding management of the municipal inventories was a real problem. Heritage practitioners were often in the unenviable position of guiding the process and attempting to provide a level of clarity on all fronts.

The State Register of Heritage Places

Assessing places for the State Register of Heritage Places was a far more complex and detailed process than assessing places for municipal inventories. The provisions of the Act protect places included in the State Register, and a great deal of detail is required to ensure that the significance of the place is accurately recorded. By contrast, places on municipal inventories are often identified with a minimum of detail. Extensive consultation with owners always occurs when places are entered on the State Register and owners can voice their objections. By contrast the public consultation required for municipal inventories is not clearly defined.

In retrospect, the years immediately following the heritage legislation produced a mixed bag of results in terms of the municipal inventory process. However, for the first time heritage was on the agenda for all local authorities. Many heritage practitioners had been actively involved in raising awareness of heritage across the state. Although the range of places identified was mixed, and the understanding of and support for the process was varied, by the year 2000 nearly all the local authorities had completed an inventory in one form or another. By this stage a number of local authorities were beginning the review process as required under the Act, and were looking at ways to provide statutory protection to all or some of the places on their municipal inventories.

Heritage protection through town planning schemes

While the *Heritage of Western Australia Act* had not made recommendations for the protection of local heritage places, by the late 1990s planning processes had developed to address this matter. The Town Planning Amendment Regulations, gazetted on 22 October 1999, gave effect to the Model Scheme Text. All local councils' zoning schemes must comply with the Model Scheme Text, except where the Minister for Planning approves a variation. The Model Scheme Text included requirements for the compilation of a heritage list, and provisions for the creation of heritage areas. The Model Scheme Text also included clear provisions for the protection of places on the Heritage List. In the case of heritage areas, protection is implemented through the creation of a local planning policy for each heritage area.

For heritage practitioners this was a welcome development because it clarified the position of the municipal inventory as a non-statutory document, but clearly identified that the heritage list must include all or some of the places on it, at the discretion of the local authority. The requirements for consultation with owners of places to be added to Heritage Lists were much more specific than the requirements for consultation under the municipal inventory process. Theoretically, an owner could find his property on an inventory without their knowledge, provided 'appropriate consultation' had been carried out to the satisfaction of the local authority. This is not the case with a heritage list attached to a town planning scheme, which required that the owner be notified.

The role played by the Heritage Council of Western Australia (HCWA) remained ambiguous. The Model Scheme Text required that the Heritage Council be advised when a place was added to a Heritage List, however the Heritage Council had no part in determining whether places were added to the Heritage List or how they would be protected. The heritage list is purely a mechanism for the local authority, which means that the contents of the list and how it is managed are at the local authority's discretion.

Municipal inventory reviews

By the year 2000 many of Western Australia's approximately 150 local authorities including approximately twenty-five in, or on the edge of, the Perth metropolitan area, had commenced reviews of their original municipal inventories (MI). In many cases the councils were also considering ways to protect some of the places listed.

Some were also reviewing their town planning schemes and adopting the Model Scheme Text. Still others however had commenced their reviews before the advent of the Model Scheme Text, and so a range of scheme provisions for heritage protection existed. In some cases there were still no provisions.

On the one hand, for the heritage professional, the situation was clearer because there were now means of protecting some of the places on municipal inventories though heritage lists in town planning schemes. On the other hand the situation had become more complicated because of the variety of different approaches to heritage taken by councils over the preceding ten years.

For the community the situation was far from clear. There was little public awareness of the planning process generally, let alone how this fitted with requirements for heritage conservation and who was responsible for what. The requirements for 'proper public consultation' for the preparation of municipal inventories and the requirements to advise owners of places proposed for heritage lists to be included on town planning schemes were not well understood.

The city of Subiaco

In the early 2000s the Subiaco Council was in the position of reviewing its municipal inventory. Subiaco's first inventory had been adopted in 1995 and contained approximately 300 places. In addition the National Trust had classified a number of places and precincts in Subiaco. There was no statutory protection for any places on the municipal inventory or places within the National Trust precincts. A small number of places had been entered in the Register of Heritage Places and were protected under the *Heritage of Western Australia Act* 1990.

As had been previously demonstrated by the Molyneux report in 1985, Subiaco was clearly an area that required consideration as a whole. An area such as Subiaco contains some individual buildings of such heritage quality that they stand out on their own. However if these were considered in isolation, the heritage value of the place would be lost. It is the grouping together of places that gives such an area its intrinsic heritage value (Spillman 1985). Furthermore, one of the major problems with the MI process was its focus on identification without consideration of management, future conservation or protection. For these reasons, the approach taken to the municipal inventory review was to firstly consider Subiaco as a series of precincts or heritage areas, and secondly to prepare it through a process that would result in heritage protection through the town planning scheme. This approach had been successful in other areas, so why not in Subiaco?

The Guildford example

Guildford, established as a town and river port in 1829, is now on the eastern outskirts of the Perth metropolitan area. It contains a range of heritage places including convict buildings, public buildings and housing stock dating from the nineteenth and early twentieth century (see Figure 12.6).

During the 1970s and 1980s Guildford was under considerable development pressure as a result of rezoning to permit high-density unit development. As a result

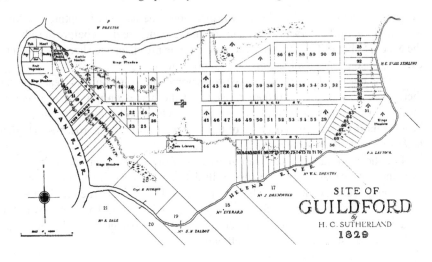

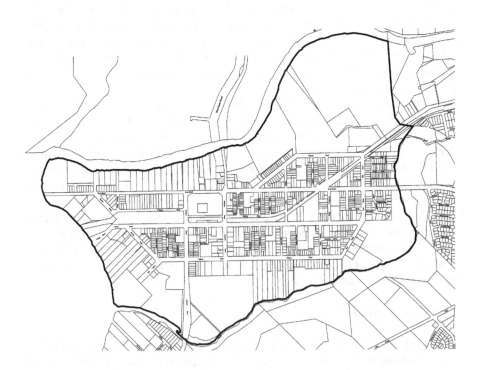

Figure 12.6   Town maps of Guildford 1829 and 2000 showing area protected as a Conservation Precinct

Source: Battye Library items approved 466C (37/13).

a number of buildings including several early homesteads were demolished. The community responded by forming a group known as the Guildford Study Group, and in 1981 prepared a report on the history and heritage of Guildford entitled 'Guildford: A Study of its Unique Character'. This report was of particular interest because it took a holistic approach and looked at the town as a series of precincts with distinct heritage character. The state government viewed the report with interest in the period that led up to the implementation of the state's heritage legislation.

In the late 1980s a partnership between the Shire of Swan and the Department of Urban Development developed a pro-forma planning strategy that resulted in an amendment to the Shire of Swan's town planning scheme to include provision for the protection of heritage precincts. The Shire (now City) of Swan's town planning scheme has included provisions for the protection of Guildford as a heritage precinct since that time.

During the 1990s the Shire of Swan went further, down-zoning the whole town to prevent further sub-division of larger lots and to protect the original town subdivision pattern that included some of the original land grants along the river. This process involved extensive community consultation with landowners, many of whom chose to support the rezoning and lose the development potential of their properties as a result. These measures were in part the result of the strength of the Guildford Association and the Guildford Historical Society, and in part of an innovative local authority.

Mount Lawley, Menora and Inglewood

Mount Lawley developed in the early twentieth century as a residential area populated largely by members of Perth's business and administrative community (Cooper and McDonald 1999). The earliest land releases occurred in the years after Federation with further development into the inter-war period. The area comprises a highly intact urban landscape of homes of varying architectural styles but with a predominance of the Federation and Californian Bungalows. Here, the Mount Lawley Society has been a voice lobbying for heritage protection since the early 1970s.

Mount Lawley, and the neighbouring areas of Menora and Inglewood, are situated under the jurisdiction of the City of Stirling. This local authority controls over 100 square kilometres of Perth's northern metropolitan area, the majority of which cannot be considered as heritage areas.

The City of Stirling had prepared and adopted its municipal inventory in 1997. This contained 640 places, but was not linked to the town planning scheme and there was not statutory mechanism for their protection. Guildford and Subiaco, the areas of Mount Lawley, Menora and Inglewood are significant as precincts or areas and not as a series of individual places. In 1999 the City of Stirling identified heritage precincts in Inglewood, Menora and Mount Lawley. Following extensive community consultation these were adopted as Character Protection Areas under the town planning scheme and design guidelines were adopted to manage development issues.

Subiaco's municipal inventory review

When the City of Subiaco set out to review its Municipal Inventory, the examples of Guildford in the early 1990s and Mount Lawley, Menora and Inglewood in the early 2000s suggested that the time had come to address the heritage of Subiaco in terms of precincts rather than as a list of individual places. It was in this context that the heritage significance of Subiaco was identified by the municipal inventory review of 2002. The review identified some 450 places or buildings as significant in their own right, and in addition identified some 2,300 places, mainly individual houses, as while not significant in their own right, contributing to the significance of nine precincts across the municipality (Minutes of Special Electors' Meeting 24 July 2002).

While there were many parallels in physical terms between Subiaco and the areas where heritage protection has been achieved, there were many differences between the separate communities. Perhaps the controversy that surrounded the release of the Molyneux Report in the 1980s, and the subsequent failure of the Subiaco Council to protect the area's heritage, should have served as a warning of what was to come.

The difference between the areas was of course most clearly marked in the area of community support. In retrospect this should have been apparent at an early stage when the council supported the concept of preparing the base material for the municipal inventory review prior to public consultation. However from the perspective of a heritage professional, this is reasonable since it is usual to develop some outline recommendations before meeting with the community. It is very important however that this is not presented as a *fait accompli* so that community consultation involves real choices, but it is also important that the professionals arrive well informed for the discussion.

The background work prepared by the heritage professionals with the city's planning staff involved looking at the ways Subiaco could be considered as a series of connected heritage areas. The answer appeared to be quite straightforward since the city's town planning scheme divided the area into precincts on the basis of historic sub-division patterns. It therefore seemed quite self-evident to consider heritage as an extra issue over the existing planning scheme structure. This meant applying management recommendations for heritage on an area-by-area basis. The result being that heritage protection is stricter in areas that are more intact so that the significance of the area becomes important, not just the significance of the individual place.

This approach for Subiaco was similar to that taken for Mount Lawley, Menora and Inglewood, however in these cases the process was not a municipal inventory review and did not involve the identification of individual properties. The existing inventory list had been used as the basis for detailed management of places within the Character Protection Areas.

The identification of individual places was extremely difficult in Subiaco due to the character of the area, the density of older housing stock and the limitations of time and budget. A street survey was carried out on foot, but the amount of detail that could be accurately recorded by such a process was limited. In retrospect, the project scope was over-ambitious and attempts to survey the whole area was one of its major failings as it raised expectations of a level of detail that could not possibly be achieved with the resources available.

Community consultation in Subiaco

The consultation programme for the municipal inventory review comprised a series of community workshops. After so many years of failing to protect heritage through statutory means, the council was keen to link the municipal inventory to the town planning scheme by adopting it as the heritage list. Since both the municipal inventory and the heritage list require public consultation, the decision was taken, in retrospect somewhat unwisely, to carry out both together under a single process. If successful this would not only result in the formal adoption of the reviewed municipal inventory, but would also establish a statutory process for managing heritage through the town planning scheme. A comprehensive package of consultation material, including not only measures for protection but also recommendations for incentives to encourage property owners to conserve their buildings, was prepared. This received a very favourable response when it was presented to the Heritage Council prior to its release.

The first workshop was held at the council offices and local politicians, community and business leaders were invited. From this point the process began to unravel. The community workshops, where the material was intended as a draft proposal for discussion, basically disintegrated. Community members appeared to be in total opposition to any concept of identifying or protecting the heritage of the area. The story filled the press and the council was inundated with calls of concern. An atmosphere of fear developed. In the midst of this a group calling itself 'Heritage Gone Mad' emerged from seemingly nowhere with demands for a public meeting. The response from the council was to call a Special Electors' Meeting in Winthrop Hall at The University of Western Australia, within the City of Subiaco's boundaries. The hope was that the size of the hall would dwarf the opposition. In the event, the scale of the opposition was completely underestimated by the city and by observers. Over time the furor died down, but for those involved it was an unnerving and to a large degree inexplicable phenomenon, quite out of scale with the issue and any perceived threat that heritage protection through the planning scheme provisions can possibly impose. Fears that having your property included on a municipal inventory or a heritage list meant that you could not change anything, could not extend, or even paint your house, were quite unfounded.

Some comparisons and perspectives

From a heritage professional's perspective, the approach taken in Subiaco was innovative, rational, balanced and intended above all to provide equity and a level of certainty for property owners. By comparison with the community of Guildford, or of Mount Lawley, Inglewood and Menora, the people of Subiaco did not have a background in heritage awareness at the grassroots level. Both the Guildford community and the Mount Lawley Society had lobbied for heritage protection in their areas from the late 1970s. In both cases consistent application of planning policies and guidelines by the council, but with community support, have been the main reasons for success.

As a senior planner with the City of Swan observed, the people of Guildford are looking for a lifestyle experience. Because of Guildford's unique character, and because heritage protection has been in place since the late 1980s, people buying into the area have a degree of certainty about what is and is not allowed in terms of development. Confidence and certainty are not developed overnight. In Guildford this has developed gradually over more than twenty years.

A town planner in private practice observed that in Western Australia there is very little appreciation of heritage other than heritage icons. He noted that generally the community accepts civic buildings such as Government House, or even Subiaco's own Regal Theatre as heritage, but it is unlikely that they see their homes, streets or neighbourhoods in the same light. He also observed that the idea of heritage areas is not well understood. Even the protection of the character of areas through design guidelines has been rejected in some areas (Nedlands) and in this context it is unlikely that controls to protect heritage, which must by definition include some level of demolition control, are likely to be accepted. That these areas remain highly desirable in terms of the real estate market indicates that factors other than the heritage character of an established area are involved.

The issue of protection of the character of an area versus the protection of a heritage area is complex. There is some overlap because heritage areas usually also have characteristics such as the styles of buildings, landscape and streetscape features that are worth protecting. The real difference however is that while character areas have qualities of visual appeal and cohesion, heritage areas are identified not only for their aesthetic characteristics, but more importantly for their historic and social values. The protection of character therefore focuses on visual factors such as building style and materials. The protection of heritage by contrast is more concerned with protecting the significant features of the area itself. This includes historic buildings, mature trees and other aspects of both the private and public domain that should be kept.

Whether areas should be referred to as character areas or heritage areas is really a matter of whether their qualities are related to historic factors or to purely visual characteristics. It is a matter of which is most appropriate to the particular situation. Subiaco's heritage values result from the early development history of the area including the subdivision pattern, the road layout, the type of houses that were built reflecting the community of people that settled there. Subiaco's heritage value results from its settlement as one of the first suburbs to develop around the city of Perth at the turn of the twentieth century.

Who defines heritage?

A prominent Subiaco real estate agent identified a major issue in the debate. He observed that the heritage field has too many players and that people are confused between the roles of the Heritage Council, the National Trust and the local authority.

The community is not able to easily understand the different interests of each authority in the heritage field. Added to this, the difference between the municipal inventory and the heritage list under the town planning scheme, was far too complicated for most people to understand the differences.

A senior town planner with the City of Subiaco, at the time of the municipal inventory review, identified the complexity of heritage administration as a major problem. The duplication of heritage identification and management through both the state heritage legislation and local planning process is too complicated. Added to this the lack of statutory responsibility of the state agency, the Heritage Council, for involvement with the administration of heritage protection at the local authority level, was evident in the Subiaco scenario.

What price heritage?

An understanding of property values is essential background to any proposal for heritage protection. It has been noted that the idea of protecting a heritage area in a country town has a very different dynamic from protecting a heritage area in an inner suburb in an economic climate of growth and rapidly rising property values. Factors that influence the value of land include its location and the zoning under the provisions of the town planning scheme. For residential land this includes the density zoning that gives the land its development potential. This, combined with the perceived market value, is a major factor determining whether heritage protection is achievable. Where zoning allows for either a change of use or increased development potential, it is unlikely that a conservation outcome will eventuate. Sometimes planning incentives can be used to tip the equation, however these must be substantial and perceived as valuable by the landowner or developer. For example, rates rebates and bonus plot ratio transfers to other sites, are real opportunities.

In areas where part of the value of the land is associated with the heritage or character of the area, such as in Guildford, there is a far higher chance of success. In Subiaco the real estate market is complex. Although the character homes are well regarded, the value of property is related to the land. As a prominent real estate confirmed, there is a demand for character homes, but the emotions that surrounded the municipal inventory review did result in a period of uncertainty in the older areas of the suburb for a period. In Subiaco many of the finest and oldest heritage homes in the area were built on main roads or near the railway and are now located on the least desirable residential streets.

Property owners' rights

Possibly the most difficult argument to address is the issue of the rights of individuals to do what they like with their properties. This was the crux of the argument against heritage protection promoted by 'Heritage Gone Mad'. In reality, the town planning process regulates a property owner's rights in terms of development options for any property. The inclusion of heritage protection as part of the planning process would therefore seem fair and just. However, in accordance with accepted heritage practice, professional assessment of a place to determine whether it has heritage significance, is concerned only with heritage values. This does not allow for consideration of land or real estate values, structural condition or any consideration of the practicality of the place for contemporary living. As a heritage professional, explaining this to a property owner is not always welcome information. Although the assessment of significance should only involve heritage values, there is a valid case for the consideration of other factors before the place is considered for protection through a town planning scheme. This is primarily the reason why a heritage area approach is important, because it ensures that all property owners in an area are subject to the same requirements.

Conclusion

In the aftermath of the controversy in Subiaco, the State Minister for Heritage established a Local Government Heritage Working Party to develop a set of common standards for the preparation of local heritage lists and heritage policies, and to develop a framework for municipal inventories. One of the major issues at state government level is that heritage, local government and planning all fall under separate portfolios. While initially the heritage legislation may not have provided adequately for the involvement of the Heritage Council with local heritage issues, it has become increasingly evident that the community does not see a clear distinction. The outcome of the Working Party was the development of a State Planning Policy for Heritage that would provide guidance to local government on the implementation of the heritage requirements under the Model Scheme Text. This included common standards for the preparation of heritage lists, a model for local planning policies for heritage areas and recommendations for incentives and education. A discussion paper was released in April 2005, but eighteen months later the policy has not been finally adopted.

From the perspective of a heritage professional, the adoption of a set of standards for heritage practice for use by local government would help to provide the clarity and certainty that is needed. However, the protection of heritage areas requires above all the realisation by communities that the history of our everyday lives, homes and neighbourhoods is important. We as practitioners need to give a clearer message that the heritage protection of an area can provide certainty about future development opportunities. Heritage protection does not mean that an area remains static or should be preserved as a museum. This is entirely contrary to contemporary heritage principles as outlined in the Burra Charter that forms the basis for professional heritage practice in Australia.

The benefits of local councils managing change in a positive manner should lead to greater opportunities for their communities. Realistic planning incentives and planning concessions in response to heritage conservation should be available and should be clearly explained. If there is something to be learned from the experience in Subiaco, it is that the heritage message should be clearer, easier to understand, more consistent and above all balanced.

References

Burke, M. J. (1987) *On the Swan: A History of the Swan District Western Australia* (UWA Press for the Swan Shire Council, Middle Swan, Western Australia).

City of Swan (2006) Local Planning Policy: Guildford Conservation Precinct; Adopted August.

Cooper, W.S and McDonald, G. (1999) *Diversity Challenge: A History of the City of Stirling* (City of Stirling WA).

Freestone, R. (1989) *Model Communities: The Garden City Movement in Australia* (Melbourne: Thomas Nelson).

Guildford Study Group (1981) 'Guildford: A Study of its Unique Character' (unpublished manuscript).

Heritage and Conservation Professionals with Sheryl Chaffer Assoc (1999) 'City of Stirling Heritage Areas Study', report prepared for the City of Stirling; March.

Marquis-Kyle, P. and Walker, M. (1992) *The Illustrated Burra Charter* (Australia ICOMOS).

Molyneux, I. (1985) *Survey of the National Estate in Subiaco* (City of Subiaco WA).

Pitt-Morison, M. and White, J. (eds.) (1979) *Western Towns and Buildings* (Nedlands: University of Western Australia Press).

Rosario, R. (1993) 'Wembley: A Geographical Study of Developing Suburban Identity in the 1920s and 1930s' (unpublished MPhil. Dissertation, The University of Western Australia).

Selwood, J. (1979) in Gentilli, J. (ed.) *Western Landscapes* (Nedlands: University of Western Australia Press).

Shire of Swan (1992) 'Guildford Conservation Policy'. Adopted by the Shire of Swan, September.

Shire of Swan and Department of Planning and Urban Development (c1992) Guildford Design and Development Guide.

Sondalini, S. '1880–1910 Geneaological Records Relating to West Australians' (State Reference Library of WA: undated).

Spillman, K. (1985) *Identity Prized: A History of Subiaco* (Subiaco: UWA Press).

Stannage, C. T., (1979) *The People of Perth: A Social History of Western Australia's Capital City* (Perth: Perth City Council).

Chapter 13

Reshaping the 'Sunburnt Country': Heritage and Cultural Politics in Contemporary Australia

William S. Logan

Australia's heritage in both its natural and cultural forms is rich, complex and unique. This reflects the diversity of the country's climate, landforms, flora and fauna as well as the juxtaposition of indigenous populations of Aborigines and Torres Strait Islanders and immigrant groups from Europe, Asia, the Middle East and Africa. Even the way we look at heritage varies, most notably between Indigenous Australians, who see no separation of natural and cultural, and the white settlers/invaders, who have tended at least until the late twentieth century to conceive of heritage in much the same manner as Europeans and white North Americans. The pattern is further complicated by the class/socio-economic dimension and perturbed in the last quarter of the twentieth century by the arrival of newcomers from countries as diverse as Vietnam, China, Somalia, Iraq and India, who cannot easily share the dominant Anglo-Celtic heritage values. Contestation over the content and significance of 'Australia's heritage' has never been far below the surface but has erupted frequently and bitterly in the last ten years as a conservative government firmly entrenched in Canberra seeks to reshape the country's value systems and self-image.

High among the issues currently facing heritage conservation in Australia is, then, the need to re-visit the notion of 'heritage', to recognise its political character and to find ways of accommodating cultural diversity, and hence inclusive conceptions of heritage, within Australian society. While we might once have thought this would be relatively easy to achieve among academics and professionals, the so-called 'History Wars' of the last decade have, however, shown this to be far from the case. The intellectual discourse about history, heritage and identity in Australia is now as divided along political lines as it has ever been. Many of the ideas underlying heritage conservation practice that were assumed in the 1980s and even into the 1990s to be relatively value-neutral, are under now challenge. Many government officials and private practitioners will soldier on, trying to ignore the cultural politics that is engulfing heritage practice, but clearly greater emphasis needs to be placed on finding ways to share, enjoy and respect the country's great cultural and natural diversity.

This chapter seeks to re-cast heritage as an element of Australian cultural politics, exploring the linkage between ideology and conservation practice. It considers some of the difficulties being experienced by the Australian heritage system that make it vulnerable to political exploitation at this particular time and that will need to be addressed over the next decade if the system is to retain its credibility and effectiveness. These include the perceptions held in different quarters that the system is, on the one hand, over-extended in respect to its planning control functions while yet being, on the other hand, still narrow in heritage content and fragmented in its efforts to provide appropriate legislation and administration. In contrast to the discussions and practical interventions currently seen at the global heritage level in the work of UNESCO and its expert advisory bodies – ICOMOS, ICCROM and IUCN[1]– little attention is being given by Australian governments to protecting the country's intangible cultural heritage. The laws and bureaucratic structures in place also militate against holistic approaches to heritage protection. While the general public seems more interested in heritage protection, paradoxically some of the bodies that have put the public case in the past, at least in the case of cultural heritage, appear to be nearing the point of collapse.

Ideological bases and interpretations of the Regulatory Framework

The Liberal Party (conservative), which has governed Australia in a coalition with the National Party (formerly Country Party) since 1996, sees much of the heritage system as a legacy of the short but tumultuous period of Australian Labor Party rule under Prime Minister Gough Whitlam (1972–5), and it has sought to put its own mark on the way Australians perceive and manage their heritage. In some respects this view of the origins of heritage protection in Australia is correct. Although an interest in architectural heritage grew in the 1940s and 1950s among related professionals and some more affluent citizens, and the natural environmental movement blossomed in the 1960s, it was in the early 1970s that the first major efforts to develop a heritage protection system were taken. Spurred on by a successful series of Green Bans in Sydney and environment-focused Black Bans in Melbourne – striking examples of union power applied to environmental causes in partnership with citizen activism – the Whitlam Government established the Hope Inquiry into the 'National Estate'. The Inquiry defined the conservation task and recommended the establishment of both the Australian Heritage Commission and the Register of the National Estate (RNE), an inventory of significant heritage places to be nominated by the Australian public. Listing on the RNE did not give a place any direct government protection, though it did enable conservationists to lobby more persuasively against development interests. Under the Australian Federal Constitution, urban development, planning and heritage management were State matters, and it was only after the Australian Government ratified the World Heritage Convention 1972 using its foreign affairs powers that it was able to intervene directly in heritage place conservation.

The first legislation in Australia with real power to protect heritage places was, therefore, at the State level – the *Historic Buildings Act 1974* in Victoria and the

---

1   United Nations Educational Scientific and Cultural Organisation (UNESCO); International Council on Monuments and Sites (ICOMOS); International Centre for the Study of the Preservation and Restoration of Cultural Property (ICCROM); International Union for the Conservation of Nature (IUCN, also known as the World Conservation Union).

New South Wales Heritage Act 1977. It has been argued that these Acts were a way for the state to divert power away from the trade unions and resident action groups and to reclaim control over the historic preservation agenda and, through this, the urban development process (Yelland 1991, 44). What is significant in these early Acts was that they established a regulatory framework for heritage conservation that separated historic buildings from the planning and management of wider heritage precincts or of areas generally, this being dealt with under the town and country (or urban and regional) planning acts in the various States. This separation was perhaps unfortunate and all heritage protection might have been better dealt with as normal planning activity under the planning act (Logan 1999). Nevetheless the focus on individual historic buildings did move into historic precincts in Victoria and New South Wales. In Victoria the work of this author's University of Melbourne Geography Honours student, Florence Davis, on South Drummond Street, was used by the Carlton Association, probably the most powerful resident action group in early 1970s Melbourne, in its submission to the Melbourne Metropolitan Strategy team in 1974. Although the Central Victorian goldfields town of Maldon had been proclaimed Australia's first 'Notable Town' by the National Trust in 1970, South Drummond Street became the first heritage precinct to be given state protection, along with the Parliament precinct in East Melbourne.

Another feature of the legislation was that it was directed principally at controlling the *re-development* of significant buildings or areas, while the business of *physically conserving* significant buildings, gardens or other properties was left to their owners. This distinction has remained in subsequent amendments and new acts. Guidelines set by government agencies are designed to help owners look after their buildings, but direct financial support from government is uncommon. The aspects of the regulatory framework dating from the early 1970s have been sporadically attacked over the subsequent 30 years as unfairly imposing restrictions and costs on private owners. Under the Liberal-National Coalition Government led by Prime Minister John Howard, with its political rhetoric highlighting freedom of choice and property rights, the criticism has reached a new level of intensity. Following a request from the Federal Government's principal adviser, the Australian Heritage Council (AHC), the Howard Government asked the Productivity Commission in 2005 to investigate the effectiveness of the systems available in Australia for protecting historic heritage places. In retrospect, it should have been no surprise that the Commission echoed the Government line, arguing that heritage controls on private property should be preceded by an agreement negotiated between the relevent authority and the individual property owner (Productivity Commission 2006).

Although there is strong community support generally for the existing level of heritage controls – certainly the Productivity Commission was unable to show any significant level of public hostility towards the controls – some criticism of the regulatory system is probably fair. The system has grown unevenly across Australia and in some states, notably New South Wales and Victoria, it could be argued that it has expanded too far and should be pulled back into line with the key principle of the Australia ICOMOS Burra Charter – that is, that the strictness of the management regime for heritage places should flow from the level of significance attributed to them. In particular some developers, architects and property owners argue that too much control is imposed on places of low level significance, and it is true that from time to time conservation campaigns for buildings or sites of marginal significance that block worthwhile attempts to create contemporary planning projects – tomorrow's heritage. One such case was the disused 1888 Sandridge Railway Bridge across the Yarra River whose preservation was seen by many to impede the coordinated development of the river as a central feature of the city. The Harold Holt Swimming Pool in the Melbourne suburb of Malvern – ironically named after the only Australian prime minister to die by drowning – was added to the State Heritage Register in 2006 in response to plans to upgrade the pool. The statement of significance refers to the pool being *inter alia* an outstanding example of brutalist architecture, a claim that does not bear comparison with the many better examples elsewhere, in Australia and the world. While recognising that it is wise to approve new developments cautiously lest views change in future about the heritage significance of places, in some cases it might be more appropriate to fully record the building or site by photographic, archival and other means and then allow them to be replaced by modern structures or simply not replaced at all.

An electoral issue at municipal, state and national levels

Governments at all levels – national, state and municipal – fear the electoral backlash that critics claim will result from the too wide and too rigid application of heritage controls. Council membership has changed dramatically in municipalities such as Ballarat in Victoria where development interests have out-manoeuvred pro-conservation councillors in local elections. But while there is often a link between electoral results and government policies towards the built heritage as part of urban planning, the electorate does not respond along neat political party lines and governments sometimes operate in unexpected ways. Recent experience in the State of Victoria seems to bear out these observations. The Liberal Party Government under Jeff Kennett that came to power in 1992 seemed to many observers to have backed the developers' argument that heritage protection had got out of hand. The Kennett Government developed a policy of urban consolidation through the encouragement of medium-density housing, immediately putting pressure on the historic building stock in inner city areas. The Minister of Planning also used a process of 'calling-in' development proposals that seemed threatened by local resistance on heritage and other grounds, and ministerial approval was given to high-rise apartment blocks overshadowing parklands and historic precincts and boulevards. Paradoxically, however, it was under Kennett's Government that Victorian municipalities were required to superimpose a heritage overlay across their local planning schemes that identified heritage buildings and to establish regulations and guidelines for them. The paradox seems to be partly explained by the fact that development interests prefer to operate within a fixed set of heritage rules rather than being caught out by new heritage inscriptions suddenly introduced after the development process has begun.

In the end the Kennett Government's record on planning and heritage matters helped bring it down in the 1999 State election. Opposition had mounted both in the inner suburbs and, especially, in the middle band of suburbs which had traditionally voted for the conservative party but where outrage over planning policies had led to the formation of a powerful 'Save Our Suburbs' organisation. The Australian Labor Party was returned to office under Premier Steve Bracks and, if heritage-minded voters expected that the new government would reassert tighter planning controls on new development and give greater support to heritage and urban character areas, they were soon disappointed. The Labor Government's actions, like those of the Kennett Government before it, have been decidedly mixed. On the one hand it has successfully (at least for the time being) resisted efforts in some affluent bayside municipalities to throw out the heritage overlay controls. On the other hand, the Bracks Government maintains the consolidation policy in its new metropolitan strategy plan, *Melbourne 2030*, and continues to use the increasingly unpopular 'call-in' powers. It has also tended to regard the National Trust and some other heritage groups and activities as reflecting only middle-class interests and, despite the high popular support for heritage protection in inner-city areas that normally vote ALP, marginal to government concerns. Whether the Save Our Suburbs group had any impact on the November 2006 Victorian State Election which saw the re-election of the Bracks government, remains to be determined.

It has been argued that a number of recent interventions by the national government into heritage issues also appear to reflect electoral concerns. Two disputes are outlined here to demonstrate a geographical coincidence between political inteventions and marginal electorates – the bitter argument over the traditional grazing practices of the so-called 'mountain cattlemen' of Victoria's Alpine region and the conflict between wind farm development and the protection of the orange-bellied parrot at Bald Hills in South Gippsland, southeast of Melbourne.

Victoria's mountain cattlemen

The mountain cattlemen case erupted in May 2005 when the Victorian State Government, acting on advice from the Federal Department of the Environment and Heritage (DEH), banned free-range cattle grazing in Victoria's Alpine National Park when the existing licences were due to expire in June 2006, on the grounds that the practice was threatening fauna and flora in the park. Grazing in the adjoining New South Wales Kosciuszko National Park had been phased out by 1972 and in the Alpine Parks of the Australian Capital Territory (ACT) by 1908 (Office of the Premier and Minister for the Environment, 2005). However, as *The Age* newspaper noted,

> Messing with a legend is a scary thing for any government. There's normally heady emotion involved, trick arguments of culture and strong public opinions. And of all Australian legends, the Man from Snowy River is surely one of the nation's most potent and enduring' (*The Age*, 25 May 2005).

Opposing lobby groups were formed, using horse parades involving actors, television stars and footballers to win media coverage. Responding to protests from the 45 affected cattlemen, and tapping into popular sentimentality towards the 'Man from Snowy River' legend portrayed in literature and film, the Howard Government, through the Federal Minister for the Environment and Heritage, Senator Ian Campbell, took the unprecedented step of using its emergency heritage powers under the *Environment Protection and Biodiversity Conservation Act 1999* (EPBC Act) to block the ban. This action was criticised for seeking to go beyond the powers granted to Canberra under the Act and for undoing the decade-long effort to establish an articulated Australian heritage system that depended on goodwill and cooperation between the Federal and State layers of government (Albanese, 2005). It was also pointed out that conserving a *place* (here, the Victorian Alpine Park) is never sufficient in itself to keep alive an intangible form of heritage (here, cattle grazing practices and associated horse-riding skills). The Victorian Minister for the Environment also pointed out that the Commonwealth could not legally force the State to renew grazing licenses even if the Alpine Park was on the National Heritage List and dismissed Senator Campbell's interventions as a 'political stunt' (Thwaites, 2005).

Eventually, in October 2005, the Federal Minister heeded the advice of the AHC, withdrew his bid for emergency listing of the Victorian Alpine park and in a media release announced a face-saving $15 million plan to create a Greater Alpine National Park straddling the state boundaries that would, subject to a favourable assessment by the Council, be considered for inclusion on the National Heritage List at a future date (Campbell, 2006b). The same media release ends with the accusation that the Victorian Government was 'hell-bent on chasing green votes at the expense of enhanced environmental management balanced with our heritage values'. But the same concern for electoral results, it can be argued, also underlies the Senator's dogged determination to promote the cattlemen's cause. The alpine Federal electorate of Eden-Monaro has been held only marginally by the Liberal Party over the last decade and is currently one of the most marginally held Liberal Party seats in Australia having been won in the 2004 Federal election by a mere 1.7 per cent of votes.

Orange-bellied parrot at Bald Hills in South Gippsland

The Bald Hills case revolves around efforts to protect the orange-bellied parrot, a small 'grass parrot' found in dwindling numbers in the coastal regions of Southeast Australia where it is estimated less than 200 remain. Using his power under the EPBC Act to intervene in a State planning matter where fauna and flora are critically endangered, the Federal Minister banned the development of a windfarm proposal. According to the Senator's media release, a government report released in April 2006 concluded that because 'almost any negative impact on the species could be sufficient to tip the balance against its continuous existence', even the 'minor predicted impacts of turbine collisions' should be prevented. (Campbell 2006a). Indeed, the report agreed that such an argument might be made but it in fact concluded that 'Our analyses suggest that such action will have extremely limited beneficial value to conservation of the parrot without addressing the much greater adverse effects that are currently operating against it', such as wildfires, disease, storm events or the genetic consequences of small population size (Smales et al, 2005, 47).

There are 20 windfarms in the orange-bellied parrot zone in south-eastern Australia and the Victorian Minister for Planning, Rob Hulls, used the same report to argue that 'not a single orange-bellied parrot was observed near the proposed Bald Hills wind farm. At best, scientist found a few historic records of sightings and a couple of potential foraging sites 10 to 35 kilometres away' (Hulls, 2006). The Opposition and media commentators have asked why, under these circumstances, the focus of the Minister's actions has been the Bald Hills proposal. According to Hulls' media release, the Federal Minister's decision is 'more about politics and keeping old promises, than it is about protecting the orange-bellied parrot' and he referred to the Liberal Party campaign against wind farms in the 2004 Federal election that helped its candidate wrest control of the seat of Macmillan from Labor. The seat has long been volatile and is now one of the Liberal Party's most marginal (estimated 3.3 per cent following a recent a recent boundary redistribution). More importantly, the case represents a conflict between two conservation aims: to reduce Australia's dependence on fossil fuels as a source of electric power and to protect a threatened bird form. The impact, if any, on the next Federal election expected by late 2007 will soon be seen.

Articulation across jurisdictions and agencies

As indicated, under Australia's constitutional arrangements planning and heritage are in the most part 'State matters'. World Heritage is an exception, coming into the national government's hands under its foreign affairs powers. Nevertheless over the last ten years very significant steps have been taken to develop an articulated system across the various jurisdictional levels. Earlier in this volume, Graeme Aplin has detailed the development of this so-called 'national integrated system' that followed the Council of Australian Governments agreement in November 1997. That this cooperative agreement was reached is particularly remarkable given that it occurred at a time when the national government was controlled by the conservative Liberal-National coalition while all six states and the two territories – the ACT and the Northern Territory (NT) – had Labor Party governments. The explanation has much to do with the foresight of Senator Robert Hill, Australia's longest serving Minister for the Environment and Heritage (1996–2001), and his successor Senator David Kemp (2001–4). But it also reflects a desire by the national government to focus its role on matters of national and world significance, to bring environmental planning controls together under a single EPBC Act and to relegate other matters to the states. The cooperative agreement also made possible for the first time in Australia the creation of a National Heritage List of places of significance to the whole nation. The states and territories for their part accepted the new system once it was recognised that places inscribed on the World Heritage List or the National Heritage List, while now falling under national legislation, would continue to be managed locally under negotiated agreements with Canberra.

Despite these advances there is a need for further strengthening of articulation within the Australian heritage system. In particular greater consistency would be advantageous in the criteria and thresholds used in inscribing places on the various registers, in the thematic approaches to building up the lists, and in ensuring that there is a strong research basis supporting the nomination and monitoring processes. There are parts of Western Australia, South Australia and New South Wales that are not well covered by heritage studies and so it remains unclear what exactly is worth protecting there. Effective articulation also requires a a commitment to incorporating heritage controls into normal planning systems across the whole country and all levels of government and a greater degree of conformity between the states and territories regarding the levels and types of controls and guidelines established by its local government authorities.

The fragmented character of Australia's heritage system not only relates to the multiplicity of levels of government, but also to the separation of different components of heritage into different government departments and instrumentalities. Thus at the national level, natural and cultural heritage are contained within a single department – the DEH, but artefacts are managed by the Department of Communication, Information Technology and the Arts (DCITA). At the state and territory level, cultural heritage protection is commonly focused in departments dealing with historic places. By this is often meant only non-Indigenous places – as if the experience of Indigenous peoples ended with the arrival of Europeans! – and Indigenous places are administered by another branch of government, as, too, are natural heritage places. It is therefore impossible to deal holistically with heritage, a serious defect, especially given the thrust of international discourse about best practice, which sees the elements needing to be brought together and dealt with by multi-disciplinary teams of administrators and practitioners.

An incomplete and unbalanced system

The focus of heritage management in Australia has been on heritage places, both cultural and natural, with an extension into cultural landscapes and underwater heritage in the 1990s. Australia's heritage is largely valued internationally because of the variety and relatively pristine condition of its physical environments and these have played a major role in shaping Australians' own sense of identity, which is reflected in folk lore, literature, art and film. In relation to historic places, the system is biased in favour of the Anglo-Celtic legacy which is commonly referred to as 'mainstream', ignoring the Indigenous heritage, which dates back 40,000 or more years and remains strong in many parts of the country, as well as the heightened cultural diversity resulting from the mass immigration programs that followed World War II. The Anglo-Celtic bias reflects power arrangements in Australia in the first half of the twentieth century when middle class city-dwellers discovered bush-walking and, in post-World War II years, when architects and associated dilettantes began to make claims for the protection of the grand nineteenth-century buildings that interested them. These latter efforts led to the formation of National Trust branches in each state and territory and to the registers of 'classified' buildings. The last 40 years has seen the filling out of the lists to reflect the more complex social and economic patterns in Australian metropolitan, rural and regional areas. This has sometimes involved a measured thematic approach, as in the DEH where each year consultants are employed to identify places of national heritage significance under selected themes. For instance, the current author led teams commissioned to work on the themes of 'Creating an Australian Democracy' and 'Australians at War' (Logan *et al.* 2003 and 2005).

In moving to widen the official registers a wider range of stakeholders have become involved and more rather than less contestation has resulted. The concept of cultural landscapes was introduced into the World Heritage system in 1992 as part of an effort to give greater credibility to World Heritage List which was seriously biased towards European heritage forms (Labadi 2005). The re-inscription of Kakadu National Park in 1992 as a cultural landscape bringing in Indigenous culture, and of Uluru–Kata Tjuṯa in 1994, showed Australian conservation practice to be at the international forefront. As long as the focus of government attention was on heritage places, management issues had a strong geographical basis and geographers played a significant role in shaping the intellectual discourse, assisting in policy formulation and implementing management plans. The growing international interest in intangible cultural heritage, on the other hand, sees the spatial element much diminished – indeed cast aside in some instances except in the most general sense. At the global level of heritage protection, as under the *World Heritage Convention 1972* and subsidiary documents, such as the World Heritage Centre's *Operational Guidelines for Implementing the Convention*, it was always possible to invoke intangible heritage values in determining the significance of a physical site, particularly under Criteria v, vi and ix. In Australia there seems to be a lack of clarity about the intangible cultural heritage concept and many members of the heritage profession and bureaucracy confuse it with the 'social' significance of places, which has become more important in the latest revision of the Australia ICOMOS *Burra Charter* (Australia ICOMOS, 1999).

UNESCO's emphasis has shifted under the current Director-General Koïchiro Matsuura (appointed 1999) towards intangible cultural heritage that is not linked to a place any more specifically defined than a nation state or major national sub-region, such as drum-playing traditions in the upland region of central Vietnam or Bagandan bark cloth-making in Uganda. There is a strong Japanese influence in this development, which links back to that country's well established 'living human treasure' concept. Starting with a program of listing 'Masterpieces of the Oral and Intangible Heritage of Humanity', UNESCO moved to a fully fledged *Convention for the Safeguarding of Intangible Heritage*, which was approved by its General Conference in 2003 and entered into force in April 2006. Article 2 of the Convention describes intangible cultural heritage as 'practices, representations, expressions, knowledge, skills' – in other words, heritage that is embodied in people rather than in inanimate objects and places. The Convention opens up a Pandora's box of difficulties, confusions and complexities, and a set of operational guidelines akin to those for the World Heritage Convention has yet to be developed (Logan, forthcoming). Those developing the guidelines have so far refused to use the World Heritage Convention's central concept of authenticity, although continuity of an intangible form must have some critical relevance to the value of the form. A new set of ethical issues also arise, since it is not possible to 'own' people in the way that we can own physical property, or to buy, sell, destroy or preserve communities of people the way we can with places and artefacts. Living culture cannot be frozen; living communities in cities or in rural areas cannot be turned into museums, at least not without the prior informed consent of the inhabitants. This means that the protection and preservation of intangible cultural heritage is very closely linked to 'cultural rights' as a form of human rights.

Australia has much significant intangible cultural heritage, such as the different forms of bushcraft developed and still practiced by Indigenous and settler communities in the 'Outback', the hybrid cultural practices of the various migrant groups, or the distinctive forms of sport created and played in Australia, such as the Australian Rules Football. But the Australian Government rejects the idea of ratifying the Intangible Convention. Its public position is that the Convention needs further work, which is indeed true; and it shows its commitment, albeit low level, to Indigenous heritage through the Maintenance of Indigenous Languages and Cultures and the media access programs administered by DCITA. It argues also that it shows commitment to intangible cultural heritage protection generally through its funding of agencies such as the National Sound and Screen Archives. But it may see the Convention as strengthening multiculturalism, a policy approach the government has been winding back since it came to office. Perhaps it fears that cultural divisions will be reinforced by any renewed emphasis on minority cultures. It is also wary about signing up to further international charters which might lead to further international interference in national sovereignty, as it perceived the UNESCO and ICOMOS interventions in the Kakadu World Heritage dispute, and its right to govern.

Heritage and the nation building agenda

The protection of heritage places, both natural and cultural, is often portrayed as a barrier to development, even to the creation of the 'heritage of the future'. The counter-argument – that, by thinking carefully about the location and design of new developments, it is usually possible to find ways to balance them with the protection of the significant heritage values – falls on deaf ears or is dismissed as representing an unnecessary level of planning control and an infringement of civil liberties. The Mirrar people's insistence on protecting their cultural heritage at Kakadu was seen as standing in the way of the development of uranium mining in the area. As Graeme Aplin has outlined (see Chapter 3), the Australian Government was incensed when the Mirrar broke the normal protocol and took their case (unsuccessfully in the long run) to UNESCO's World Heritage Committee. Jane Lydon and Tracy Ireland (2005, 15) point out, however, that 'For Aboriginal people, representing hundreds of small linguistic and cultural entities across the continent, and excluded from citizenship until 1967, no allegiance to a national framework can be assumed'. They see an Aboriginal 'strategy of refusal' that fractures consensual narratives of the nation, with dramatic political statements such as the establishment of a Tent Embassy on the lawns outside Australia's Parliament House in Canberra (*ibid.* 17). If this is the case, then the notion of 'shared' heritage needs to be replaced by one of parallel and perhaps irreconcilable voices.

Labor governments in the various states often demonstrate strong neo-liberal approaches to government, putting development projects ahead of heritage protection. For instance, the highly significant Indigenous rock art on the Burrup Peninsula on the Dampier Archipelago of Western Australia's Pilbara region has been regarded by the Labor Government in that state as standing in the way of the development of BHP-Billiton's iron-ore shipping facilities and construction of a new Woodside Petroleum plant for processing its NW Shelf natural gas reserves. A Draft Management Plan 2006–2016 has now been drawn up by the WA Department of Environment and Conservation and released for public comment in mid-2006 (Western Australian Department of Environment and Conservation, 2006). It proposes an arrangement under which 62 per cent of the peninsula (essentially the non-industrial land) will be jointly managed by the traditional custodians – the Ngarda-ngarli – and the Department. Meanwhile the AHC released its assessment report in early October 2006 recommending 874 square kilometres of the archipelago, including 100 square kilometres of the Burrup, be put on the National Heritage List. This has placed the Federal Minister for the Environment and Heritage in a difficult situation and he has announced that he will delay making a decision on whether to act on the Council's recommendation. There is a feeling that the Federal Government might wish heritage issues to disappear from the political agenda, and, now that it is in control of both houses of the national parliament, it has moved to revise the EPBC Act 1999 in order to rein in the AHC.

On the other hand governments are very happy to use cultural heritage when it suits their development agendas. Heritage is the basis of Australia's tourism industry, with the Sydney Opera House, the Great Barrier Reef and Uluru–Kata Tjuta among the top sites visited by international and domestic tourists. The first mentioned is currently under UNESCO consideration for World Heritage listing, while the last two are already World Heritage listed. While economically advantageous, the commercialisation of heritage is not without its dangers in terms of maintaining site authenticity and, ultimately, the sustainability of both the heritage and the tourism trade dependent on it (Timothy and Prideaux, 2004). But there is another, ideological agenda – nation-building – that exploits heritage and that can lead, in its most extreme forms, to dangerous incursions into democratic institutions and the human rights of people, especially minority groups. Cultural heritage, in particular, can be and is used to manipulate people and governments commonly use it to shape public opinion or to try to weld disparate ethnic and social groups into more cohesive and harmonious national entities. All of these manipulative activities may be benign if they promote tolerant states and societies based on human rights; but in too many cases governments have used selective versions of the 'national cultural heritage' to force minority groups to adopt the dominant culture, effectively wiping out their own cultural identity (Logan, forthcoming).

Many critics are now arguing that multiculturalism is in a state of crisis in democratic societies around the world where governments are retreating from open commitment to cultural diversity, emphasizing instead security and integration (Isin and Turner, 2006). This reflects the new governmental focus on both neo-liberal economic and social arrangements and on the 'war on terror'. Governments espousing neo-liberalism give primacy to business interests, especially global, and quarantine some things from public scrutiny, notably major development projects. They oversee and encourage fundamental economic and social restructuring, with, in Australia particularly, little vision of the longterm future for the workforce. The role of the state has been distorted, especially in relation to its defence of human rights and democratic institutions and practices. In Australia, critics argue, there has been a reduction in civil liberties in the pursuit of 'national security' on the one hand, but, on the other hand, an emphasis on the 'human right' of individuals to do what one wants with property.

A perception of increased contestation within Australian society has clearly led over the last ten years to the political response on the part of the Howard Government of downplaying the nation's cultural diversity and of seeking to impose a new sense of 'Australian-ness', especially by determining the interpretations of 'the nation' to be taught in schools and universities, displayed in museums and protected under heritage legislation. Federal funding requires all schools, for instance, to fly the Australian flag and display a 'values' poster showing the iconic Simpson and his donkey, while history teachers and museum directors are exhorted to concentrate on celebrating the achievements of post-1788 settlement. There is already a substantial literature on the so-called 'History Wars' in the academic and popular press, and the controversy is kept alive by recent prime ministerial addresses (McIntyre and Clark, 2003; Grattan, 2006). Such efforts are characterised by some critics as a nostalgic quest to recapture an imaginary, homogeneous culture of the 1960s, but others see an insidious curtailing of the rights of Indigenous peoples and recent ethnic immigrant groups to maintain their own cultures (Casey 2004; but see also McIntyre 2006). Following a serious outbreak of violence in the Aboriginal town of Wadeye in Western Australia in May 2006, a Menzies Research Centre report was released by the Federal Department of Education, with the Minister of Education's endorsement, that advocated the removal of indigenous culture from the primary school curriculum ion the grounds that it was preventing Aboriginal children from progressing in their education (Khadem, 2006). It also recommended closing schools in remote communities if they are considered to be economically unviable.

Historian Marilyn Lake (2005) has concluded that 'Foreign battlefields have displaced frontier wars as sites of memory'.

> Who cares whether Aboriginal people were dispossessed by British settlement or that colonial history was marred by massacres? Real Australian history begins with Gallipoli, when Australian men joined the first Australian Imperial Force to fight overseas – not so much, it seems, for God and Empire as old memorials still somewhat embarrassingly insist – but for modern Australian freedom. And the men kept fighting for freedom during World War II, in Malaysia, Korea and Vietnam, in the Gulf and now Iraq. This is history to Howard and it is getting a lot of air-play.

The Prime Minister and other Government members and officials have indeed made numerous visits to battlefields in recent years, setting up memorials and encouraging pilgrimage tourism. There has been particular government interest in Anzac Cove at Gallipoli, Turkey, the site of a disastrous encounter with the Turkish army in World War I that has acquired iconic status not only for the huge Australian sacrifice of lives but also because of the way in which Australians were seen to have displayed key qualities of their national character, notably mateship and determination, and as a place where they realised that their future had to be one of independence from Britain. The research conducted by Deakin University for the 'Australians at War' thematic study for the DEH showed that this is the war-related site most valued by Australians, far outranking any war-related site within Australia (Logan et al, 2005). Australia can only influence the conservation of Anzac Cove by using extra-territorial means and negotiations with the Turkish Government have taken place to find a way to inscribe Anzac Cove on the Australian National Heritage List. These are unsuccessful at this stage, no doubt set back by the controversy that erupted in 2005 when archaeologists and heritage groups claimed that the Federal Government and Turkish authorities had allowed fragile relics at Gallipoli to be disturbed by the widening of the access road that would enable larger numbers of tourists to visit the place.

The Government is clearly using a selection of cultural heritage items as the core around which it is seeking to reshape the nation. The new National Heritage List, which currently stands at 34 inscribed places, reflects the government's approach. At this stage the List is not overly ideological in its construction, however, and includes the Eureka Stockade Gardens with its links to political disturbance and radical change, and Aboriginal sites at Lake Condah, Brewarrina and Hermannsburg alongside places that celebrate European discovery and settlement of the continent, such as Dirk Hartog's 1616 Landing Site, Cook's landing place at Kurnell Peninsula and the Royal Exhibition Buildings. This is because the process is based on public nominations, although the Minister approves listings and is able to use this power to prevent places being added where they are seen as interfering with development projects (eg. Burrup?) or opposing the Government's nation-building efforts (eg. Aboriginal massacre sites or refugee detention centres). The intention of the AHC Chairman, Tom Harley, has been to make the List mean something to the general public and this has had the effect of taking responsibility away from heritage professionals. Thus the List includes sites such as the Glenrowan Historic Precinct associated with Ned Kelly, which does not have a high degree of authenticity in the Venice Charter sense. Pushing the populist approach to an extreme, the Government announced the inscription of the Melbourne Cricket Ground on the National Heritage List during the 2005 Boxing Day Test cricket match and the inscription of the Flemington Race Course on Melbourne Cup Day 2006. There is no problem with making the List more socially inclusive; there is, however, a fine irony in the fact that the heritage significance of places such as the MCG and Flemington depends on intangible values, yet the Government resolutely resists signing up to UNESCO's Intangible Convention.

A decade ago, at the height of the Kennett years in Victoria, conservation architect Nigel Lewis called for greater consistency and impartiality in the maintenance of conservation standards. 'There is an urgent need to redefine our objectives and create a culture that sees conservation as a long-term process, free from the influence of sudden political swings and the vagaries of public taste' (Lewis, 1997, 59). The situation is even less convivial today and the need is even greater for heritage protection to be seen as a long-term social and cultural objective that ought to be separated from short-term electoral interests. Whether heritage can be competely separated from long-term ideological interests is, however, a moot point, heritage being a subjective concept and inevitably contested, as Tunbridge and Ashworth (1996) and many others have observed. But it will be troubling, indeed, if the government pushes further with its history wars attack and the List loses balance. The bill currently before parliament to emasculate further the Australian Heritage Council does not bode well, and the Minister's decision on Burrup is something of a litmus test. Furthermore, a separate list of Commonwealth Government-owned properties is also kept under the EPBC Act and an indication of the current Federal Government's less than whole-hearted support for heritage conservation is seen in the fact that, since the heritage provisions of the Act came into force, new Commonwealth properties have been added to the Commonwealth Heritage List at the rate of only one a year. The government's favoured approach is to sell off properties rather than to have them assessed for their heritage values and maintain them where they have heritage significance.

The failure of public action and a special role for geographers

The way forward is surely to define heritage widely and inclusively and to avoid narrow interpretations of Australia's history and heritage based on the views of the 'dominant' social and political group. Minority voices, whether Indigenous or immigrant, must be incorporated in the formation of Australian identity. Governments should resist the temptation to manipulate Australian 'mainstream' attitudes by recourse to sensationalist misrepresentation of opposing views of the nation's past and current culture, but should listen to all groups in the community and encourage them to participate in heritage identification, management and interpretation. However, if we accept that heritage identification and preservation is essentially political, then it will always be necessary to put pressure on governments in order to achieve the goals of inclusivity and balance. Unfortunately at the moment in Australia there is little intellectual resistance to neo-liberal government, to the negative aspects of economic and cultural globalisation, or to the view that private property rights are unlimited. This is in part the result of the environment of insecurity and self-censorship that has been created over the last decade. Paradoxically, therefore, we face a situation in which there appears to be strong community support for heritage conservation in both rural and urban areas and for the maintenance of high quality museums and other cultural collections, and yet there seems to be a collapse of public action in the heritage field. We are unable to depend on some of the usual sources of pressure on policy makers, especially the National Trust, which has its own internal problems. Resident action does not exist on the 1970s scale and householders now seem more concerned about interest rates than their neighbourhood environments. Community and professional groups have faced funding cuts where seen to be too critical of government policies. The media have shown little critical interest in the Productivity Commission report or the bill to emasculate the AHC. The overwhelming task for conservation groups is, therefore, to find new ways of getting the message across to the public, of mobilising community concern and of influencing the decision-makers.

There is a role for universities in this context in articulating alternative views and helping to balance the public debate. The business management model that now pervades Australian universities does not help to achieve this (Logan, 2007); nor does political interference in university research funding (McIntyre, 2005). But universities have a traditional responsibility to provide intellectual leadership, and within universities, a special onus falls on geographers to use their unique set of synthesizing skills and interests to challenge the neo-liberal social and economic development approach and to focus on the key issue of reciprocity – that is, of showing how and why human rights should continue to be supported but of also articulating the view that there is a reciprocal set of duties that humans have towards each other and their physical and cultural environment. The geography agenda insofar as it relates to heritage issues thus spans both empirical and theoretical studies and includes the study of the power relationships that impact on the landscape. It should draw out the meanings and ironies of our 'Sunburnt Country', embrace the intangible values of places and help to achieve more holistic and culturally sensitive approaches to environmental understanding and protection.

References

Albanese, A. [Shadow Minister for the Environment and Heritage] (2005) *Campbell careless and incompetent with Australia's heritage* (Media release, Canberra, 17 June).

Australia ICOMOS (1999) *The Burra Charter: The Australia ICOMOS Charter for Places of Cultural Significance* (Australia ICOMOS Secretariat, Deakin University, Burwood, Victoria).

Campbell, I. [Senator, Minister for the Environment and Heritage] (2006a) *Wind farm report warns of risk to Orange-bellied Parrot* (Media Release CO66/06, Canberra, 5 April).

Campbell, I.[Senator, Minister for the Environment and Heritage] (2006b) *Decision on Alpine National Park Nomination* (Media Release CO298/05, Canberra, 14 October).

Casey, D. (2004) *The National Museum of Australia – A Clash of the Cultures*, Centre for Public Culture and Ideas seminar, Griffith University, Brisbane, 2 June.

Davison, G. and McConville, C. (1991) *A Heritage Handbook* (Sydney: Allen & Unwin).

Grattan, M. (2006) 'PM claims victory in culture wars', *The Age*, 26 January.

Hulls, R. [Victorian Minister for Planning] (2006) *Wind farm decision overturned due to endangered bird* ('The World Today' ABC Radio program, 6 April).

Isin, E. F. and Turner, B. S. (2006) *Reclaiming Citizenship: The Contradictions between Citizenship and Human Rights*, Paper presented to Centre for Citizenship and Human Rights, Deakin University, 6 November.

Khadem, N. (2006) 'Stop Aboriginal culture lessons, report says', *The Age*, 30 May.

Labadi, S. (2005) 'A review of the Global Strategy for a balanced, representative and credible World Heritage List 1994–2004', *Conservation and Management of Archaeological Sites*, vol. 7, pp. 89–102.

Lake, M. (2005) 'The Howard History of Australia', *The Age*, 20 August.

Lewis, N. (1997) 'Urban conservation and urban consolidation – 20 years progress?', *Historic Environment*, vol. 13, no. 1, pp. 54–62.

Logan, W. S. (1999) *Zoning and Land-Use Codes for Historic Preservation: the Melbourne Example*, Paper presented to the Economics of Heritage: UNESCO Conference on Adaptive Re-use of Historic Properties in Asia and the Pacific, Melaka and Pinang, Malaysia, May 1999.

Logan, W. (2007), 'Heritage education at universities', in M.-T. Albert and S. Gauer-Lietz (eds), *Heritage Education: Capacity Building in Heritage Management* (Frankfurt am Main: IKO – Verlag für Interkulturelle Kommunikation, & Cottbus: Brandenburg Technical University). pp. 64–9.

Logan, W. S. (forthcoming), 'Closing Pandora's Box: Human Rights Conundrums in Cultural Heritage Protection', in H. Silverman and D. Ruggles Fairchild (eds), *Cultural Heritage and Human Rights* (New York: Springer).

Logan, W. S., Beaumont, J., Ziino, B., Smith, A., and Donkin, J. (2005) *Australians at War. Report to the Australian Department of the Environment and Heritage* (Melbourne: Cultural Heritage Centre for Asia and the Pacific, Deakin University).

Logan, W. S., Stokes, G. M., Long C. D. and Gillen, K. (2003) *Creating an Australian Democracy. Report to the Australian Heritage Commission* (Melbourne: Cultural Heritage Centre for Asia and the Pacific, Deakin University).

Lydon, J. and Ireland, T. (eds.) (2005) *Object Lessons: Archaeology and Heritage in Australia* (Melbourne: Australian Scholarly Press).

McIntyre, D. (2006) 'The National Museum of Australia and public discourse: the role of public policies in the nation's cultural debates', *Museum International* (UNESCO), vol. 58, issue 4, December 2006, pp. 13–20.

McIntyre, S. (2005) 'Research floored by full Nelson', *The Age*, 16 November.

McIntyre, S. and Clark, A. (2003) *The History Wars* (Melbourne: Melbourne University Press).

Office of the Premier and Minister for the Environment [Victoria] (2005) *High Country grazing continues outside National Park* (Media release, Melbourne, 24 May).

Productivity Commission (2006) *Conservation of Australia's Historic Heritage Places, Final Report* (Melbourne: Productivity Commission).

Smales, I., Muir, S. and Meredith, C. (2005) *Modelled Cumulative Impacts on the Orange-bellied Parrot of Wind Farms across the Species' Range in South-Eastern Australia, Report for Department of Environment and Heritage* (Melbourne: Biosis Research Pty Ltd).

Thwaites, J. [Victorian Minister for the Environment] (2005) *Campbell stunt on Alpine grazing exposed* (Media release, 10 June).

Timothy, D. J. and. Prideaux, B. (2004) 'Issues in heritage and culture in the Asia Pacific region', *Asia Pacific Journal of Tourism Research*, vol. 9, no. 3, pp. 213–23.

Tunbridge, J. and Ashworth, G. J. (1996) *Disonnant Heritage: the Management of the Past as a Resource in Conflict* (Chichester: Wiley).

Western Australian Department of Environment and Conservation (2006) *Proposed Burrup Peninsula Conservation Reserve Draft management Plan 2006–2016* (Perth: Department of Environment and Conservation).

World Heritage Centre (2005, rev. edn) *Operational Guidelines for the Implementation of the World Heritage Convention* (Paris: UNESCO).

Yelland, S. (1991) 'Heritage legislation in perspective', in Davison and McConville (eds). *A Heritage Handbook* (Sydney: Allen & Unwin).

Index

Printed in the United States
by Baker & Taylor Publisher Services